Alaska's Predators

WOLF TRACKS IN THE MUD.
PHOTO COURTESY OF THE US FISH AND WILDLIFE SERVICE.

Alaska's Predators

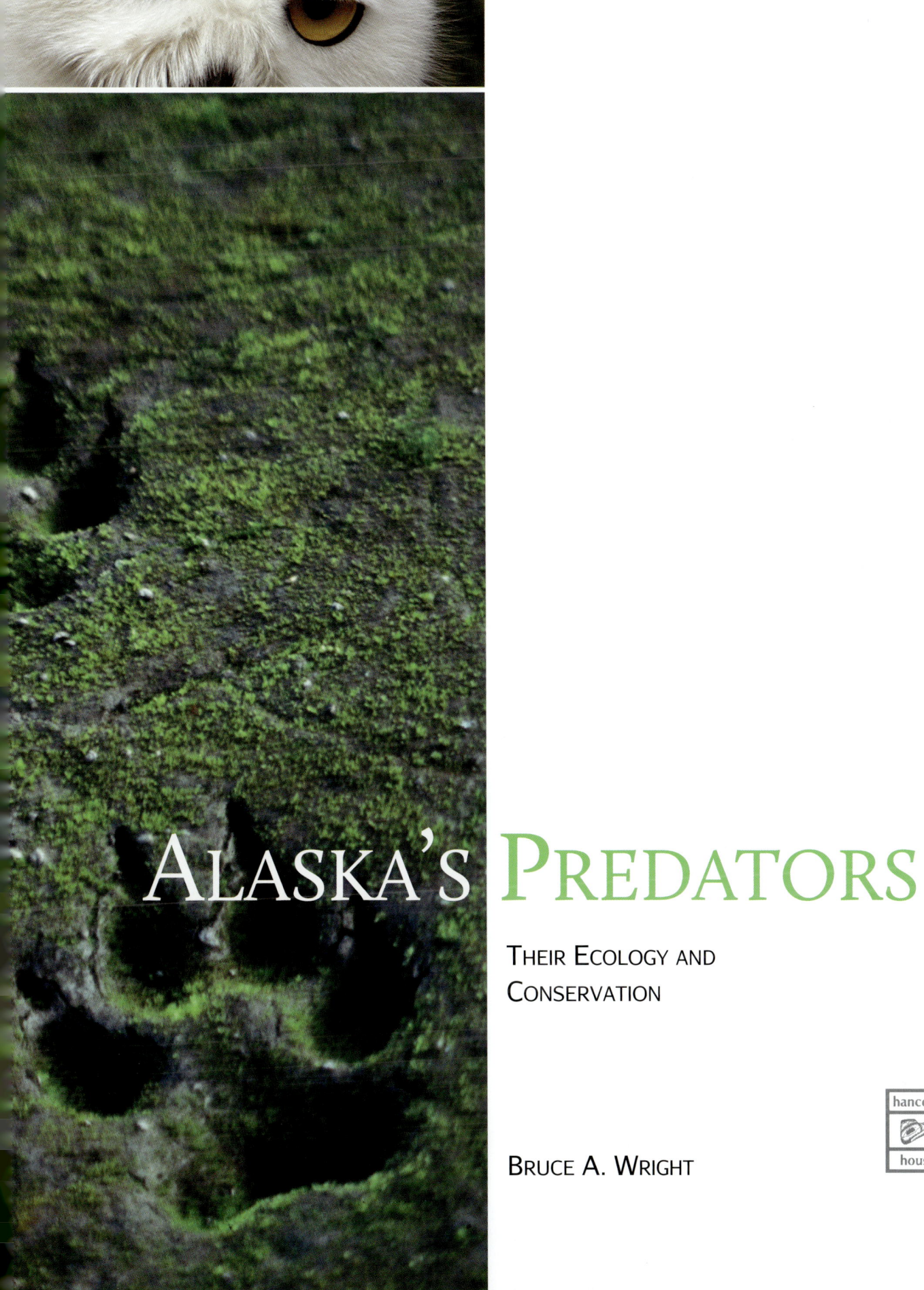

Their Ecology and Conservation

Bruce A. Wright

ISBN-10 0-88839-622-8
ISBN-13 978-0-88839-622-8
Copyright © 2011 Bruce A. Wright

Cataloging in Publication Data

Wright, Bruce A., 1952–
  Alaska's northern predators : their ecology and conservation / Bruce A. Wright.

  Includes bibliographical references and index.
  ISBN 978-0-88839-622-8

  1. Predatory animals—Alaska. 2. Predation (Biology)—Alaska. I. Title.

QL758.W75 2011   591.5'309798   C2007-902174-3

All rights reserved. No part of this publication may be reproduced, stored in a retrieval system or transmitted, in any form or by any means, electronic, mechanical, photocopying, recording, or otherwise, without the prior written permission of Hancock House Publishers.

Printed in South Korea — PACOM

Editor: Nancy Miller, Theresa Laviolette
Production: Mia Hancock
Cover Design: Mia Hancock

Published simultaneously in Canada and the United States by

HANCOCK HOUSE PUBLISHERS LTD.
19313 Zero Avenue, Surrey, B.C. Canada V3S 9R9
(604) 538-1114 Fax (604) 538-2262

HANCOCK HOUSE PUBLISHERS
1431 Harrison Avenue, Blaine, WA U.S.A. 98230-5005
(604) 538-1114 Fax (604) 538-2262

www.hancockhouse.com
sales@hancockhouse.com

Table of Contents

Preface / 7
Foreword / 8
Acknowledgements / 10
Introduction / 11
    Alaska, A Land of Superlatives / 11
    Ecological Roles of Predators / 12
    Ecosystems, Food Webs and Indicator Species / 12
    Keystone Species, Top-Down and Bottom-Up Control / 14
    Predator Control / 15
    Regime Shifts and the Northeast Pacific / 16
    Optimal Foraging Theory and Relationships of Predators and Prey / 17
    Home Range and Territory / 18
    Landscape Ecology and Gene Flow / 19
    Patches and Hot Spots / 20
    Niche and Interspecific Competition / 21
    Conservation / 22
The Species — Land, Air and Sea / 23
    Terrestrial Predators
        Arctic Fox / 25
        Wolf / 28
        Black Bear / 32
        Brown Bear / 36
        Lynx / 43
        Mink / 45
        Marten / 48
        Wolverine / 51
    Avian Predators
        Bald Eagle / 57
        Gyrfalcon / 62
        Great Horned Owl / 65
        Snowy Owl / 69
    Marine Predators
        Sunflower Star / 73
        Giant Pacific Octopus / 76
        Salmon Shark / 79
        Pacific Sleeper Shark / 82
        Pacific Halibut / 86
        Sea Otter / 90
        Steller's Sea Lion / 94
        Polar Bear / 98
        Orca / 103
        Great White Shark / 109
Glossary / 112
Information Sources / 115
About the Author / 116
Index / 117

Dedication

This book is dedicated to Roy Kreighbaum and George Wright, who bestowed an appreciation and respect for nature.

Preface

In this book you will read about the animals I have come to know, through research and personal experiences, as the predators of Alaska.

When my daughter, Sadie, was only six years old and she and I were on a midnight clamming adventure, we were surrounded by a pack of howling and growling wolves. The wolves were but feet away and numbered more than 20, but they allowed us to retreat across the clam beach of Bridget Cove and to our car.

On the pristine Alaska marine waters I have experienced orcas that approached my skiff, rubbed up alongside and seemed to eye me with curiosity. When watching sea lions near Poundstone Rock in lower Lynn Canal, one of the sea lions decided to jump into the boat. At the time, Sadie, by then 15 years old, was filming them with a video camera, and although the camera only caught the action on audio, I shooed the beast from the skiff.

I have rescued injured bald eagles, encountered wolverines on several occasions and have been kept at a distance from raptor nests by protective snowy owls, northern goshawks, and great horned owls.

One time I was in the wilderness, approximately 70 miles from the nearest village, Illiamna, on the Alaska Peninsula. Four grizzly bears had been moving across the tundra towards my camp in a dense alder patch. As the day passed and the sun was nearing the horizon, I crested a small hill to find myself only 35 feet (10 m) from the four huge grizzlies. I yelled, "Hey! Get out of here!" Three of the bears stood on their hind legs and the fourth froze on all fours. Two of the giants started to shake their heads. The bears seemed to be trying to figure out what the human was doing, why a person was in their wilderness home, and if they should attack or flee. Again I yelled, "Go on! Get out of here!" The three standing bears were swaying, as if the Alaska Peninsula winds were rocking them back and forth; this was their way of sizing things up. As suddenly as the confrontation had begun, it ended. The bears turned and ran with great resolve, bouncing off each other, taking turns to glance over their shoulders to make sure the human was not in pursuit.

Predators are charismatic animals that most people enjoy seeing. Visitors to Alaska want to see a bear or wolf moving across the tundra in search of its next meal. Watching salmon sharks leaping out of the water, chasing their prey, ignites some primal emotion in the viewer. Seeing a bald eagle swoop down to snatch an unsuspecting fish from the water is fascinating to onlookers, and the raptor's flight and sight capabilities are impressive and awe-inspiring.

For this book, I have selected representative predators, species I have studied and which I find especially interesting. Many other Alaska predators exist, such as beluga whale, coyote, cougar, Dall's porpoise, dragon flies, fisher, grey whale, harbor seal, humpback whale, northern goshawk, peregrine falcon, red fox, river otter, salmon, sperm whale, weasel, and white-sided dolphin. There wasn't the space to include them all, but perhaps the other Alaska predators will be incorporated into a subsequent book.

Three brown bears, a sow with her two cubs, in an Alaska stream. *Photo by John Harvey.*

FOREWORD

Alaska, best described by the state motto the Last Frontier, is known for its spectacular scenery and wildlife. Among the most sought-after wildlife species for viewing are the predators. People flock to Denali National Park and other remote locations in hopes of seeing bears and wolves. Marine wildlife tours strive to bring their clients close to orcas, sea lions and sea otters. In Alaska you have the chance of seeing 400-pound salmon sharks and bald eagles pursuing their salmon prey. In *Alaska's Predators* you will learn about these intriguing animals, and you will gain insight into how they live and operate in Alaska's vast wilderness.

The author, Bruce Wright, has worked with predatory animals throughout Alaska. During his work and travels in Alaska he has heard tour guides and others tell tall tales of bald eagles weighing 50 pounds with the ability of carrying off a small child, of wolves being as likely to take healthy animals as those that are sickly, and other misleading and inaccurate stories. This book will be useful to tour guides, giving basic and interesting information to share; and this book will provide information for the wildlife viewer, tourist or Alaska resident, as well, so that they may better understand the lives of these complicated creatures. Bruce has done an excellent job providing information to a wide variety of people with an interest in wildlife, including students and first-time wilderness visitors. *Alaska's Predators* presents information about the most sought-after predators and also some of the lesser-known predator species. *Alaska's Predators* is for people who are knowledgeable about wildlife, as well as for those who have little or no science background but who have a great interest in the wildlife of Alaska. This book will provide insight into how these beautiful animals manage to survive, and highlight the natural history that makes these species so interesting.

Alaska's Predators will provide a catalyst for discussing the importance of predators and lead to a greater understanding and protection of these species and their environment. Though top-end predators receive much attention in the mainstream media, their depiction is usually of a bloodthirsty, malevolent entity intent on hunting and killing everything in its path. Television, in particular, capitalizes on the biological imperative of these animals, sensationalizing and dramatizing their mode of survival, for profit. However, top predators are very important for maintaining diverse and healthy ecosystems, and they perform this service to the benefit of many other species, including humans. Accordingly, there is a great need for people to understand predators and to disseminate the truth about these species, to dispel erroneous information, and work to promote laws to protect predators for the common good. *Alaska's Predators* will serve to stimulate further study for a better understanding of predators and particularly of their conservation.

JIM AYERS
VICE PRESIDENT, OCEANIA

• • •

Predators are important elements in ecosystems as they influence and are influenced by their prey. Many of the ecosystems in Alaska retain much of their wilderness character, and they still have healthy populations of "charismatic megafauna" like bears, wolves, eagles, and sharks. Nevertheless, these and other predators are sensitive to changes in the ecosystems, and large changes in their populations signal shifts in ecosystem conditions. Predators are often in the news in Alaska. Whether due to direct contact with people (like interactions between hikers and bears, or between commercial fishermen and seals or killer whales) or due to management proposals (such as reducing predators, like wolves, to increase populations of prey, like moose), predators often foster lively debate. Introductions of predators, such as mink, to some Prince William Sound islands, marten to islands in Southeast Alaska, and fox to the Aleutian Islands, has resulted in dramatic drops in bird populations. The foxes systematically are being removed from the Aleutian Islands, and bird populations are responding well. However, mink and marten are still affecting their prey and ecosystems on islands in Southcentral and Southeast Alaska.

Climate change is exerting itself in Alaska, and scientists are predicting dramatic effects from the warming climate. Along with warming ocean and terrestrial ecosystems, we expect to see increases in the establishment of invasive species. Scientists are particularly worried about Atlantic salmon, (*Salmo salar*), Northern pike (*Esox lucius*), yellow perch (*Perca flavescens*), Chinese mitten crab (*Eriocheir sinensis*), green crab (*Carcinus maenas*), signal crayfish (*Pacifastacus leniusculus*), New Zealand mudsnail (*Potamopyrgus antipodarum*), zebra mussel (*Dreissena polymorpha*), hydrilla (*Hydrilla verticillata*), dotted duckweed (*Landoltia punctata*), purple loosestrife (*Lythrum salicaria*), Eurasian water-milfoil (*Myriophyllum spicatum*), Japanese knotweed (*Polygonum cuspidatum*), saltmarsh cordgrass (*Spartina alterniflora*), swollen bladderwort (*Utricularia inflate*), and reed canarygrass (*Phalaris arundinacea*). Invasive diseases likely will be unstoppable and cause significant changes to Alaska's plants, animals, and ecosystems.

To better inform people about some of Alaska's predators, Bruce Wright has assembled an informative array of material on predators in both marine and terrestrial food webs in Alaska. He is well qualified for this project because of his training in biology and his experience observing and studying wildlife in various parts of this vast region. Over the past three decades, Bruce has studied several top-level predators in Alaska, including eagles, bears, sharks and other marine and terrestrial species.

This is not only the first book on Alaska's predators, but this is also one of the very few books on predators published worldwide. Bruce did an admirable job of summarizing the most important findings in predator biology, and the book also includes a concise general account of predators in Alaska including arctic fox, bald eagle, black bear, brown bear, great horned owl, gyrfalcon, Pacific halibut, Pacific sleeper shark, killer whale, lynx, marten, mink, octopus, polar bear, salmon shark, Steller's sea lion, sea otter, snowy owl, sunflower star, wolverine and wolf. In addition, each species account includes information about the species' size, color, speed, longevity, reproduction, social structure, distribution, movements and migration, habitats used, prey, predators, predatory characteristics, current status, and ecology and conservation. This new approach allows the reader to compare the predators, their ways of finding food, avoiding other predators, and reproducing.

Alaska's Predators' strength, however, comes, from the presentation of information about ecological and wildlife management concepts, including: ecological roles of predators, ecosystems, food webs, indicator species, keystone species, top-down and bottom-up population control, predator control, regime shifts in the Northeast Pacific, optimal-foraging theory, relationships of predators and prey, home range and territory, landscape ecology and gene flow, patches and hot spots of prey abundance, niche and interspecific competition, and conservation. The beauty of *Alaska's Predators* is that the information will be useful to the non-scientist and scientist alike. This is truly a book that will be used in the university classroom and also will be appropriate on the coffee table in any home.

Although Bruce is a working scientist, he is aware of the need to make summary scientific information available to broader audiences, and he has compiled information in this book that will be of broad appeal. The book is written in a style that will make the scientific material accessible to all kinds of readers. Read straight through, the text will give an excellent overview of the ecology of predators in different ecosystems. Individual species accounts are useful as a reference to people with broadly ranging interests.

G. VERNON BYRD
SUPERVISORY WILDLIFE BIOLOGIST
ALASKA MARITIME NATIONAL WILDLIFE REFUGE
US FISH AND WILDLIFE SERVICE

Acknowledgements

My oldest daughter, Sadie Wright, and I have shared many experiences with Alaska's wildlife, and we both studied these animals in university and work with them professionally in Alaska; I thank Sadie for taking so much of her valuable time to edit and make recommended changes to this book. Thanks go to those people who reviewed species for the book: Jose Castro, Chris Krenz, Craig Matkin, Steve Peterson, Eva Saulitis, and David Scheel. I also would like to thank Catherine E. Taylor for her critical review and constructive comments. I thank Gus Mills, author of *African Predators,* for his useful comments and guidance. I would also like to show my appreciation for all the scientists, researchers, educators, and resource managers who work so hard as advocates for wildlife and wild places.

Introduction

As you read through *Alaska's Predators* you will gain an understanding of these magnificent animals, their function in Alaska's ecosystems, and how and why we should protect them. Below are explanations of many of the ecological concepts that will help you to understand how predators survive. I recommend that you review these now and refer to them again after reading about your favorite predators on the following pages. Many of these ecological concepts also are discussed in context within the species descriptions. I have tried to minimize the use of technical terminology, but terms are defined when they are used. The glossary at the end of the book also should be helpful in this respect.

Alaska, A Land of Superlatives

Alaska, with 586,000 square miles (1,517,733 km^2), is one-fifth the size of the lower 48 states combined; if Alaska were divided into two states, Texas would be the third-largest state. Alaska has 33,904 miles (54,563 km) of coastline (twice that of the continental US), more than 3,000,000 lakes larger than 20 acres (8.1 ha), more than 3,000 rivers, 63 percent of the nation's wetlands, 39 mountain ranges, 17 of the 20 highest peaks in North America, 1,800 islands and more than 100,000 glaciers.

Temperatures range from –80ºF (–62ºC) in winter to 100ºF (38ºC) in summer. Alaska has four climate zones: maritime, transition, continental and arctic. The temperature range generally is greater farther from the coast. Permafrost covers most of the northern third of Alaska.

The arctic zone receives less than six inches (15 cm) precipitation per year with amounts increasing to the south and nearer the coasts. The continental zone averages 12 inches (30 cm) per year, the south side of the Alaska Range receives nearly 60 inches (152 cm) and the maritime zone receives annual precipitation in excess of 200 inches (508 cm). Snowfall makes up a large portion of the total annual precipitation.

The greatest recorded annual precipitation occurred at MacLeod Harbor in Southeast Alaska with 332 inches (843 cm).

A MOUNTAIN SCENE WITH ALASKA COTTON GRASS GROWING NEAR BEAVER PONDS.
Photo by Bruce Wright.

The diversity of Alaska's geography and climate provides a wide range of habitat types for Alaska's predators. Alaska's low human population density, about one person per square mile, reduces human-caused effects to Alaska's wildlife.

Polar bear.
Photo courtesy of John Gomes and the Alaska Zoo.

Ecological Roles of Predators

Different predators have different ecological roles and effects on the ecosystem. Wolves and coyotes may appear very similar, but their hunting strategies, pack dynamics, and prey are quite different. Wolves tend to take large, old and sick animals while coyotes often take smaller animals without regard to their condition. Removals of some large predator, through predator-control programs actually may result in large increases in smaller predator populations. When wolf populations are targeted and reduced, one can expect a dramatic increase in coyote populations. This may be a result of increased prey availability or a reduction of predation on coyotes by wolves. This is known as the "meso-scale predator release phenomenon." Likewise, when coyote populations are reduced, fox populations prosper. Healthy predator populations of wolves, coyotes and foxes may require less total prey for this suite of predators.

Armed with new information on predator–prey populations, and growing public sentiment to protect predators, people have supported laws and management strategies to protect these important and interesting species. Even resource managers, who often support predator-control programs, are attempting to manage wildlife and fisheries under a new paradigm — ecosystem management based on ecosystem research. Ecosystem management is a departure from single-species management, in that many of the biotic and abiotic components of an ecosystem, and their relationships, are considered in management decisions. This holistic approach is necessary for managing, protecting and restoring our ecosystems. It is important that managers consider the role of predators in the ecosystem.

Wolverine.
Photo courtesy of John Gomes and the Alaska Zoo.

Ecosystems, Food Webs and Indicator Species

An ecosystem is a plant and animal community together with its environment, functioning as a unit. An ecosystem may be as small as the community of organisms that live under a small rock or as large as the Amazon Basin tropical rainforest. The earth is an ecosystem connected by weather and atmosphere, oceans, and plant and animal populations. Ecosystems are in flux and always changing, either from natural forces or from human activities. Each time that a part of an ecosystem system is altered (e.g., something as simple as a single, fallen tree) the entire system is affected. The single tree crashing to the ground usually does not affect humans, who live and perceive on a scale that prevents understanding the consequences of a fallen tree. The beetle that lived under the tree's bark, or the robin that nested in the tree, is more likely to perceive the scale of a single, fallen tree. However, when a forest is destroyed, such as when trees are clear-cut in a watershed, people may perceive silt filling streams and rivers, the fish in that stream may perish, and local wildlife may leave the area. These landscape-size changes are more perceptible to humans.

The linkages of the ecosystem are through the movement of energy and nutrients that occurs through a chain of organisms: producers (plants), consumers (herbivores and predators), and decomposers (fungi and other microscopic organisms). All three of these living ecosystem components are necessary for a system to function. The conversion of the sun's rays to stored energy by plants is very inefficient: approximately 1 percent of the radiant energy from the sun is converted to plant biomass. This limits the amount of energy available to the rest of the ecosystem. The consumer group consists of herbivores, which feed on plants, and predators, which feed on other animals. The links between the producers, herbivores, predators and decomposers are referred to as the "food chain." In a food chain an animal incorporates only about 10 percent of the energy it ingests. The majority of energy in a food item is used up in body maintenance, movement and energy conversion. Of course, this is a simplistic view of ecosystems and the food chain. In reality, the links in the system are complex and interwoven, forming a web of connections and interactions — a "food web" of life.

GYRFALCON. *Photo courtesy of Mark Robb.*

A BALD EAGLE IN SEARCH OF PREY.
Photo courtesy of John Gomes.

Predators are a necessary component of the ecosystem, as important as plants (producers), consumers (herbivores), and decomposers. If any of these ecosystem components are eliminated, or even reduced, the system may change drastically. The consequences of such a change can be severe, even catastrophic. Ecosystems that experience such dramatic shifts may show signs of less species diversity, lower productivity, or domination by lower-level smaller species.

Food chains usually contain links of six species or less. The longest chains exist in aquatic systems. Since so much energy is used to maintain each subsequent link in the food chain, the amount of available energy decreases rapidly at every trophic (energy or food) level, and each level supports fewer individuals than does the one before. This results in a pyramid of numbers, or a food pyramid with many organisms at the bottom (producers) and few at the top (predators). An ecosystem that is in trouble will likely magnify the discrepancy further up the food pyramid. Scientists can monitor the ecosystem by monitoring the species at the top of the food pyramid. Scientists monitor these "indicator species," often predators, to detect effects from human activities such as heavy fishing pressure.

Indicator species also can be used to detect increasing contaminant levels (e.g., DDT) and how these chemicals move through the food web. This may be more important for people now than in the 1960s, when eagles and other predators in eastern North America indicated that the environment was contaminated with high levels of DDT. Today there are more than 85,000 synthetic chemicals registered in the United States, and scientists do not know the environmental or human health implications associated with most of them. For example, these chemicals may be linked to increases in breast cancer. Just 50 years ago women had a one in 22 chance of developing breast cancer in their lifetime. Today they have a better than one in eight chance. Indicator species likely will be useful in the future for helping people to detect effects from hazardous synthetic chemicals. This could help save the lives of wildlife and people alike.

Keystone Species, Top-Down and Bottom-Up Control

One way that scientists categorize animals is by what they eat: *herbivores* eat plant material, *carnivores* eat other animals and *omnivores* eat both animals and plants. Certain species play a pivotal role in affecting the entire system, a role that is disproportionate to their numbers and biomass. These are called "keystone species," and they may be herbivores, omnivores or carnivores. Keystone species are identifiable in that they have low biomass but large effects on the community structure. A keystone species can be responsible for the structure and integrity of an ecosystem. Many keystone species are predators.

The omnivores and carnivores, both groups within the larger group referred to as predators, are at the top of the food pyramid. If these species are keystones species in an ecosystem and have a significant influence on the lower levels of the food web, this ecosystem is referred to as having "top-down" control. An Alaskan example of top-down control of a system is demonstrated by orca predation on sea otters. Sea otters are very good at preying on sea urchins. When the sea otters are removed from the system, for example, eaten by the orcas, the sea urchin population can explode and overgraze the macro marine algae (large seaweeds), which in turn will alter the habitat for many of the smaller near-shore species. Sea otters and orcas, both predators, are keystone species because they influence a community to a greater extent than expected from their biomass.

This large orca is capable of taking prey many times its size.
Photo courtesy of John Harvey.

"Bottom-up" control is when the ecosystem structure is dependent upon factors such as nutrient concentrations or prey availability from lower trophic levels. An Alaskan example of bottom-up control is the reduction of phytoplankton (small, usually microscopic plants) populations due to low nutrient availability (possibly the element iron), which reduces the food availability for microscopic herbivores, reducing the food availability for consumers. This ultimately affects top predators.

Scientists attempt to understand the ecosystem so that they can determine what is happening, why the system is changing, and predict what it will do in the future. If scientists could determine the future abundance of an important species, salmon for example, they could recommend harvest levels to optimize production while protecting the spawning population (escapement). Tracking the microscopic plants (phytoplankton) and even intermediate species can be problematic and expensive. However, tracking top predators, especially if they are keystone species, is easier and very revealing of the workings of the ecosystem. As you read about the 21 predator species selected for this book, think about insights scientists and resource managers may gain from understanding population numbers and trends of these Alaska predators.

Predator Control

Predators may compete with people for some resources. In Alaska much attention has been given to controlling bears and wolves that are preying upon moose and caribou. Alaskan officials have sanctioned programs to reduce some predator species in certain regions to promote increases in large ungulates (e.g., moose and caribou), with the ultimate goal of providing more food for rural subsistence people and better hunting opportunities for people living in cities. These predator-control programs do not take into consideration the evolutionary advantages of the cycles of predator and prey populations, but instead try to promote unnatural stable population trends.

Some people believe predators are in direct competition with humans and must be controlled (numbers reduced) or eliminated. However, predator–prey relationships are complex. For example, in the mountains of New York State a healthy wolf population would be expected to take approximately 4,000 white-tailed deer per year. In this particular situation the wolf population may indirectly improve habitat from overgrazing by the burgeoning deer populations. However, the wolves have been removed from this ecosystem, the area is overgrazed by the high deer population, and nearly 18,000 deer die each year of starvation during the harsh winters. As seen in this case, predators can have indirect effects on prey population dynamics. By reducing older, less viable females, they promote overall reproduction of a herd with more young that would have increased levels of survival. Instead, humans eliminated the wolves of New York years ago and the deer suffer the consequences.

Wolves have been reintroduced to Yellowstone National Park with many unexpected consequences. One of these consequences was the improvement of fish habitat and fish populations. At first, it may not be obvious why fish would benefit from the return of wolves to a system, but the wolves used the streamsides to ambush elk. The elk soon learned to avoid the riparian zones and the vegetation grew back from being over-grazed by the elk. The recovered streamside habitat provided more cover and food for the fish and the fish population rebounded.

> If the biota, in the course of eons, has built something we like but do not understand, then who but a fool would discard seemingly useless parts? To keep every cog and wheel is the first precaution of intelligent tinkering.
> —Aldo Leopold

Some wildlife managers have forgotten that the characteristics of the wildlife we most value evolved in part from the relationship between predators and their prey. The evolution of the four-chambered stomach in ungulates developed as a result of a need to eat fast in the open and chew later in the safety of cover. Animals have keen senses of hearing, smell and vision because of their relationship with predators. Scientists are finding that predation, particularly predation by large carnivores is a necessary component of all healthy ecosystems. Study after study has shown that predator loss leads to biodiversity loss and a reduction in the quality of ecosystems.

STELLER'S SEA LIONS HAULED OUT ON A COBBLE-COVERED ALASKA BEACH. *PHOTO COURTESY OF THE US FISH AND WILDLIFE SERVICE.*

Regime Shifts and the Northeast Pacific

Dramatic changes in the species composition of ecosystems have been documented throughout the world. During these biological regime shifts dominant species may become less abundant and new keystone species may regulate the system. Such a regime shift occurred in the Gulf of Alaska during the late 1970s. During this period there were declines in some predators (seals and marine birds), and increases in others (salmon and sharks). Global warming, resulting in an increase in ocean water temperatures, may have promulgated the regime shift. During the late 1970s regime shift, scientists noted declines of important high-fat, small, schooling fish, called "forage fish." Shrimp populations also declined to a small percentage of their former abundance. A small predator, a copepod, is now abundant where it once was rare. The copepods produce a waxy, nearly indigestible compound that may be reducing production in some predators and not affecting others, thus driving a bottom-up regime shift. Seabird diets reflected these changes, which resulted in dramatic declines of the reproduction rates of many seabirds. Three large predatory species benefited from the regime shift: fish-eating resident orcas, Pacific sleeper sharks and salmon sharks.

Different keystone species have taken over, which may exert a top-down predation effect on the system. Many cycles and changes in ecosystems are normal and expected. Resource managers and policy makers should use adaptive management strategies (changing harvest levels, periods of harvests, use of protected areas, etc.) to adjust to these changes, based on sound ecosystem management philosophy.

Another consequence of global warming may be the reduced capability of water to hold oxygen. Colder water holds more oxygen and phytoplankton may produce more omega 3 fatty acids, an important building block of cell membranes. Increased seawater temperatures could easily drive regime shifts.

SUNFLOWER STAR. *PHOTO COURTESY OF JOHN HARVEY.*

OPTIMAL FORAGING THEORY AND RELATIONSHIPS OF PREDATORS AND PREY

Ecologists theorize, develop complex mathematical models, and test hypotheses in an attempt to understand the complexities of predator–prey relationships. Some of the earliest observations of predator–prey relationships focused on tracking the success of fishermen. Fishermen gravitate to the areas where fish are abundant and the fishing effort is optimal, where they can catch the most fish for the least amount of effort. This is referred to as "optimal foraging strategy" and applies to many predator–prey interactions. You probably have seen animals use optimal foraging strategies: hummingbirds incorporate the high-energy foods from feeders put out by people, coyotes feed where lots of rabbits are found, some sparrows hang out at fast-food restaurants looking for an easy meal, and ravens and bears frequent garbage dumps where edible scraps are abundant.

Often predator and prey develop a coupled relationship. An analysis of historical North American fur-trapping data reveals a coupled relationship between hare populations and one of their primary predators, the lynx. The eight- to ten-year cycles are obvious; the prey populations lead the population rise and fall, followed by the rise and fall of the lynx population. When prey is abundant, the predators respond with higher reproduction and higher survival rates of young. When prey populations fall due to lack of food (overgrazing), an individual predator will starve, move to where prey is abundant, or switch to other prey. Lynx are so closely coupled to hares that, soon after the hare population declines, the lynx population suffers large mortalities from starvation and reduced reproduction. This dramatic response of predators-to-prey population cycles is termed "boom and bust."

LYNX. *PHOTO COURTESY OF JOHN GOMES AND THE ALASKA ZOO.*

Buffers to the boom-and-bust cycles of prey and their predators include mechanisms of prey defense and avoidance, prey escape, armor, mimicry and refuge from predators. All of these mechanisms increase the complexities of the predator–prey relationship.

Just one of the techniques, prey defense, is used and adapted to several different situations. Prey defense may include camouflage, such as is used by ptarmigan and hare. They turn white during the winter to blend in with the snow, and mottled with brown during the summer to blend in with underbrush and exposed earth. Some prey avoid predators by being nocturnal or, as with some moths, avoid predation by detecting the sounds made by their primary predators, bats. A well-known defense used by skunks and other animals is to emit strong and/or noxious smells.

Scientists continually learn of new mechanisms of prey defense and avoidance. The multitude of predator-avoidance mechanisms increases the complexity of the ecosystem making the job of managing these populations extremely difficult for biologists.

Home Range and Territory

The "home range" of an individual animal is the area it uses throughout its lifetime. This home range encompasses all of the resources necessary to the survival and reproduction of the individual animal: food, mates, dens, etc., for all seasons. Different species of animals have different home-range sizes. The area of habitat or landscape used by a predator is dependent upon the species of predator, prey species and density, habitat type, time of year, etc. The "territory" is the part of the home range that is defended against intrusion by other members of the same species. All predators have a home range, but not all predators defend a territory.

Eastern Pacific humpback whales have a large home range. They winter in Hawaiian or Mexican waters where they breed and give birth; during the summer the whales travel to Alaskan waters to feed. Humpback whales don't defend a territory, but males do compete for females in their winter range and try to exclude other males.

Predators use a variety of techniques to defend a territory. Some predators use display and visual cues. The white head and tail of the bald eagle is visible from long distances and is useful in signaling other eagles to stay away. Bears may defend a small fishing site by displaying their size, sometimes turning sideways towards their adversary so as to look larger. Vocalizations may be used to defend a territory. Wolves howl to, among other things, exclude non-pack members from a territory. Great horned owls also use vocalizations to warn off other owls. Some predators use scent marking for

WOLVERINE.
PHOTO COURTESY OF JOHN GOMES AND THE ALASKA ZOO.

DURING THE SPRING THIS BROWN BEAR CAME DOWN FROM ITS WINTER DEN TO FEED ON THE SEDGES ON AN ADMIRALTY ISLAND BEACH.
PHOTO BY BRUCE WRIGHT.

WESTERN ALASKA MOUNTAINS IN THE FALL.
PHOTO BY BRUCE WRIGHT.

defending their territories. Wolves and foxes will urinate and defecate in specific locations, perhaps on the trailheads to valleys or along ridges, to warn others to stay away. Wolverine, marten and mink mark their territories with powerful scent-gland excretions. These techniques are used instead of physical aggression to minimize potentially lethal injuries.

However, physical aggression is sometimes necessary in territorial defense to lend credence to territorial signposts and warnings. Bald eagles, following a chase, may clasp talons and free-fall, cartwheeling with their wings spread out. The opponents usually release before hitting the ground, but there are occasions when they crash to their deaths. Bears and sea lions can be seriously injured during their physical encounters, and many have large scars that indicate the seriousness of their wounds. Maintaining a territory can be dangerous and use lots of valuable energy, but the benefits can include controlling good feeding and nesting sites or exclusive mating rights.

LANDSCAPE ECOLOGY AND GENE FLOW

Certain species of animals have tremendously large home ranges, spanning mountain ranges and vast tundra plains. These species require huge expanses of space and topographic variation in order to survive and reproduce. These vast areas used by some animals are called "landscape."

The main plant community or habitat type in a landscape is called the "matrix," and the smaller unique habitat types within the matrix are called "patches" (see Patches and Hot Spots below). In the marine environment, patches can be hot spots of a school of forage fish, and in the terrestrial environment a hot spot or food patch may be where berries or salmon are abundant.

MARTEN MAY TAKE REFUGE IN TREES OR USE TREES IN THEIR SEARCH FOR PREY. PHOTO COURTESY OF JOHN GROSS.

Animals use corridors within the matrix as avenues for traveling in their landscape. If corridors are disrupted by a natural phenomenon (a mountain range or desert) or man-made phenomenon (a road or city), segments of a species population may become isolated. This has become a serious problem in developed areas of the world. One of the primary concerns here is the lack of sexual interaction between these isolated populations, resulting in reduced or no gene flow. Isolated components of a population may suffer increased mortalities due to disease and other factors.

PATCHES AND HOT SPOTS

Prey patchiness is a characteristic that has evolved to help prey find mates and food while avoiding predators. However, some predator species have evolved to benefit from the patchiness of prey. For example, in a terrestrial system a patch can be a high-density lemming population; in a marine ecosystem the patches can be schools of high-energy forage fish such as herring. Predators have evolved to become very good at finding patches and exploiting them. Predators use search mechanisms to locate high-density patches or hot spots of prey. If the predator does not locate the patches of high abundance, the predator's reproductive success is likely to be reduced, or it may even starve. Species that can return to annual hot spots (e.g., forage fish spawning events) have a better chance of surviving and they may produce more offspring. A good example of this is the annual concentration of thousands of bald eagles along the Chilkat River near Haines, Alaska. The eagles return to this hot spot each year in the winter to feast on late-run salmon.

A BROWN BEAR FEASTING ON SALMON. PHOTO BY SCOTT GENDE.

Salmon and salmon shark numbers have increased dramatically in the northeast Gulf of Alaska since the 1977–79 regime shift. The sharks appear at salmon streams just as the salmon arrive. For a few months the sharks will intercept the waves of returning salmon, and then after the salmon have moved up the stream to spawn, the sharks disperse to other hot spots. Bears and eagles also exploit salmon runs to take advantage of the abundant energy and nutrients that will promote the predators' survival. Many pred-

SALMON SHARK. *PHOTO BY BRUCE WRIGHT.*

GREAT HORNED OWL. *PHOTO COURTESY OF MARK ROBB.*

ators depend upon a series of hot spots to increase reproductive output and survival. Following these predators from one hot spot to the next is like connecting the dots on a map.

Many patch locations are determined by habitat type and landscape. In a terrestrial ecosystem the location of a prey patch or hot spot may be its association with surface water or sun exposure. These types of ecological factors are controlled by the geology and geography of the landscape. Patches in the ocean may also be controlled by geology and geography. Upwellings bring nutrient-rich waters to the surface (the euphotic zone) creating the right ingredients for a phytoplankton bloom, then a zooplankton bloom, and ultimately attracting forage fish and their predators. Upwelling areas, such as Lower Cook Inlet and the waters surrounding the Barren Islands, result from the coastal current running into an underwater rock shelf. The waters in this upwelling are churned and mixed creating the conditions for a hot spot. Next time you find yourself at a hot spot try to determine why it is there. Is it because of the geology and geography of the landscape or seascape?

NICHE AND INTERSPECIFIC COMPETITION

Each predator has its own way of finding food, avoiding predation, reproducing and surviving. This is sometimes referred to as their "niche." You will learn from the following species accounts that many species of predators occupy the same habitat. The predators avoid conflicts and competing with other species (interspecific competition) by what is referred to as "niche separation." For example, wolves and wolverines may occupy the same territory, but the wolves favor taking large prey found at lower elevations while wolverines usually scavenge the wolf kills or take small prey, such as ptarmigan and arctic hare.

Sea lions and sharks also share the same habitat and their diets overlap; both species eat salmon, squid, rockfish, pollock, herring, capelin, sablefish, sculpin and tomcod. These two species appear to have an extensive niche overlap and compete with each other. This may help explain why shark populations do well when sea lion populations suffer.

Conservation

There are many good reasons for conserving our natural resources, including the inherent value of nature and the many benefits nature offers to humans. Many of the factors affecting wildlife and wilderness are discussed in this book: contaminants, loss of habitat, over-fishing, invasive species, clear-cut logging and global warming. Some of the tools used by conservation biologists also are discussed in the book, including establishment of marine reserves, protection of predators, and monitoring of indicator species. The task is huge, but we know future generations will benefit from our work.

A FEMALE SEA LION WITH HER PUP.
PHOTO COURTESY OF US FISH AND WILDLIFE SERVICE.

The Species

The predator species included in this book are presented by the habitat associated with their predatory behavior: land, air and sea. The terrestrial species are presented first and include arctic fox, wolf, black bear, brown bear, lynx, mink, marten and wolverine, and are organized by taxonomic family groups, Canidae, Ursidae, Felidae and Mustelidae, respectively.

The birds: bald eagle, gyrfalcon, great horned owl, and snowy owl, are organized by their taxonomic family groups, Accipitidae, Falconidae and Strigidae, respectively.

The marine species include two invertebrates: sunflower star in the Asteriidae family, and giant Pacific octopus in the Octopodiae family; and seven vertebrates: salmon shark in the Lamnidae family, Pacific sleeper shark in the Dalatiidae family, Pacific halibut in the Pleuronectidae family, sea otter in the Mustelidae family, Steller's sea lion in the Otariidae family, polar bear in the Ursidae family, and orca in the Delphinidae family.

Many of these species use more than one of the habitat types. For example, bald eagles depend upon terrestrial habitat for nesting and resting, but most of Alaska's bald eagles obtain their food from aquatic habitat. Black and brown bears are mostly terrestrial species, but polar bears spend much of their time on or in the ocean, so in this book they are considered to be a marine species.

The Alaska marine and terrestrial ecosystems are dynamic and complex beyond the human ability to readily comprehend. Predators are central to stabilizing these systems and they are important indicators of the health and observed changes to the ecosystem. Predators are not to be feared, but to be understood. Take the time to read about some of the predators in Alaska, and learn about their habitats and how they play an important role in the ecosystem. Similar basic information is provided for each species: size, color, speed, longevity, reproduction, social structure, distribution, movements and migration, habitat, prey, predators, predatory characteristics, current status, and ecology and conservation. This should help you to compare these animals and to understand the similarities and differences that lead to their success. I hope you will learn to appreciate predators for all that they do and are.

> There is a pleasure in the pathless woods,
> There is a rapture on the lonely shore,
> There is society, where none intrudes,
> By the deep Sea, and music in its roar.
> I love not Man the less, but Nature more.
>
> —Lord Byron

Killer Whales. *Photo by Bruce Wright.*

A top view of polar bear swimming in the ocean. *Photo courtesy of National Oceanic and Atmospheric Administration.*

24   Alaska's Predators

(BELOW) WHITE COLOR PHASE ARCTIC FOX DURING THE WINTER.
PHOTO COURTESY OF US FISH AND WILDLIFE SERVICE.

Terrestrial Predators

Arctic Fox

Arctic fox (*Alopex lagopus*), polar fox or white fox, are in the Canidae family, the group that includes dogs and wolves.

SIZE: Arctic fox are small, but taller than a domestic cat, and weigh 6–10 pounds (2.6–4.4 kg). They average 43 inches (1.1 m) in length including the 15-inch (38 cm) long tail. Compared to the closely related red fox, the arctic fox has short legs and short ears, and its dense winter fur gives it a stocky appearance.

COLOR: During the winter the arctic fox has a thick white coat. In April and May the winter coat is shed for a shorter, less dense, two-tone brown coat. It changes back to its thick winter coat in September and October. A percentage of arctic foxes have a bluish-gray coat year around. Of the two color phases that occur in Alaska, the blue phase is more common on the Aleutian and Pribilof islands. The blue color phase arctic fox was favored by the fur industry and introduced on the Aleutian Island chain in the 1930s. The other color phase, white, is more common in the northern populations. Both color phases may occur in the same litter.

SPEED: Arctic fox can reach speeds of 30 miles per hour (48 km/hr) but rarely do so. They seem to prefer to maintain a slower steady pace in their never-ending search for food. The faster pace would result in overheating.

LONGEVITY: Arctic fox can live over 10 years, but the average life span for arctic fox that reach adulthood is around three years in the wild.

REPRODUCTION: In early spring, arctic foxes form mating pairs. Mating follows a playful courtship of chasing and play fighting. During the 52-day pregnancy the pair locates a den in an area with adequate prey to raise the young, or whelps. Den sites are selected carefully; dens are located on south-facing slopes of gravel hills, as they must have good drainage to avoid flooding. The dens may be used repeatedly. The whelps are born in late May or early June. The average litter size is seven, with up to 22 whelps reported. This is the largest known litter size for any wild mammal in the world.

ARCTIC FOX WITH THICK WINTER BLUISH-GRAY FUR.
PHOTO COURTESY OF JOHN GOMES AND THE ALASKA ZOO.

GRAY COLOR PHASE ARCTIC FOX DURING THE SUMMER.
PHOTO COURTESY OF US FISH AND WILDLIFE SERVICE.

At birth the whelps are blind and helpless but are covered with hair. Each whelp weighs only about two ounces (57 grams). The male does all the hunting during the last week of pregnancy and the first five to six weeks after the birth of the whelps. The whelps are weaned after five to six weeks when the female joins the male in hunting and providing for the growing offspring. The whelps begin leaving the den and hunting on their own when they are three months old. By September the family begins separating, beginning their solitary winter travels searching for prey. Overwinter survival for whelps is low, but for those that survive, they are sexually mature at nine to ten months.

Arctic fox.
Photos courtesy of Dr. Natalya Tatarenkova.

SOCIAL STRUCTURE: Arctic fox will exclude other fox, including red fox, from their denning sites, but otherwise tolerate their own kind. Sometimes many individuals will group and feed together on a large cache of food such as a caribou carcass, beached whale or walrus. Large fox concentrations sometimes lead to rabies outbreaks. These occur mainly in western Alaskan fox populations.

DISTRIBUTION: Arctic fox are found in treeless coastal areas of Alaska from the Aleutian Islands north to Point Barrow and across the Arctic.

Movements and Migration: When the lemming populations crash, the arctic fox will move to other areas in search of food, switch to other prey or starve. During these movements arctic fox may travel hundreds of miles (>500 km).

Habitat: Arctic fox are restricted to treeless, open habitat.

Prey: Most arctic fox are dependent upon lemming populations for their survival. Lemmings experience dramatic fluctuations in numbers. The fluctuations include periods where lemmings may number in the thousands per acre (5,000 lemmings per ha) or the population may "crash" to an average of less than one per acre (less than two lemmings per ha). These population crashes may be due to lack of food (over-grazing), disease or predation, but numerous factors probably apply. To supplement their diet, arctic fox will eat eggs, birds, snowshoe hare, arctic hare, ground squirrels, beached fish, berries and carrion. During the winter some arctic fox follow polar bears to scavenge scraps from the bears' meals.

Predatory Characteristics: Arctic fox will use the pounce-and-pin technique to catch lemmings. Lemmings do not hibernate and are active all winter. Foxes pursue them during the winter, too. Some arctic fox will follow wolves and human hunters to eat scraps left behind. Others will venture onto the shifting ice pack to follow polar bears. The foxes will eat leftover seal scraps from the polar bear kills, or seek out and kill seal pups hidden in snow caves near breathing holes.

Predators: Wolves will eat foxes if they can catch them or find one caught in a trap. Eagles may catch whelps at the den site, and barren-ground grizzly bears and wolves dig them from the den. Fox and wolf populations usually co-exist with little conflict. However, when coyote populations are high fox populations may suffer, probably from competition and/or predation. Coyotes appear not to do well when wolf populations are high, again probably due to wolves killing the coyotes. This three-way relationship, wolves–coyotes–foxes, is muddied further, depending upon the prey populations' dynamics. Perhaps the predator–prey competition between wolf–coyote and coyote–fox pairs fuels these antagonistic relationships. Most arctic foxes live far enough north that they usually do not interact with coyotes.

Current Status: Arctic fox populations are healthy in Alaska. They are being removed systematically, mostly by trapping, from many Aleutian Islands (where they were introduced by people working in the fur industry) in an attempt to return the islands to their original state and enhance breeding bird populations. Arctic fox population densities vary depending upon prey abundance.

Ecology/Conservation: Arctic fox reproduction rates, number of surviving whelps and adults are dependent upon the abundance of their food supply. Most arctic fox populations depend upon lemming populations, which are notoriously cyclic, fluctuating from year to year.

Arctic fox with a seabird.
Photo courtesy of US Fish and Wildlife Service.

Arctic fox pelts on fur stretchers.
Photo by Bruce Wright.

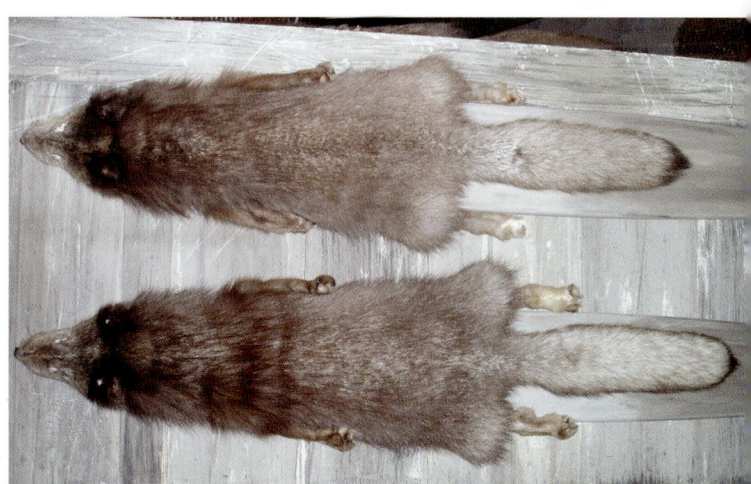

When the lemming populations are high, fox reproduction rates are high and over-winter survival may be high. But when the lemming population crashes, the arctic fox will move to other areas in search of food, switch to other prey or starve. This boom-and-bust situation applies to other northern prey–predator groups, including lynx–hare, wolf–caribou, snowy owl–lemming, and wolverine–ptarmigan.

Winter arctic fox pelts are very attractive and historically have been very valuable. From 1910 to the 1920s, when fur prices were at their peak, fox were introduced to several of the Aleutian Islands. The fox were harvested for their fur, but when fur prices dropped they were left to their own devices. These foxes exerted a huge toll on the nesting seabirds, and Aleutian Canada geese were ill prepared to deal with the introduced predators. Ground and burrow-nesting seabirds were most at risk, but the cliff-nesting species also suffered from fox predation. Within 50 years several species of the isolated islands' nesting-bird populations had plummeted. Aleutian Canada geese were extirpated on many islands. Nutrient transport by marine birds from the ocean (bird doo-doo) was reduced, resulting in changes in abundance and species composition of the islands' plant communities.

In 1977 the Aleutian Canada Goose Recovery Team was formed to protect and promote recovery of the few remaining geese (about 1000 individuals). A huge effort by the federal government resulted in eradication of foxes and reintroductions of goose breeding pairs to many of the islands.

Wildlife viewing and ecotourism has become popular in the Arctic, and the arctic fox is one species that tourists have a fair opportunity of seeing. As human activity in the Arctic increases, including wildlife viewing, fox populations may increase when they gain access to garbage dumps and other artificial food sources. Attracting and feeding these canines can increase the risk of transmission of diseases, such as rabies, or artificially raise the local population of foxes. Feeding wild animals should be avoided. Some arctic fox even have learned to follow researchers to bird nesting sites. After the researcher leaves the area, the fox may eat the eggs or chicks.

Wolf

Wolves (*Canis lupus*) are in the Canidae family, which includes foxes, coyotes and dogs. In Alaska two subspecies are recognized, wolves in Southeast Alaska, and wolves in the remainder of the state.

Size: Male wolves can exceed 200 pounds (90 kg), though this is rare. The average adult male weighs about 85–145 pounds (39–66 kg) and is about 26–32 inches (66–81 cm) tall at the shoulder. Females are slightly smaller than males. Wolves that inhabit the Southeast Alaska archipelago are generally smaller than the wolves that live in interior Alaska.

Photo courtesy of US Fish and Wildlife Service.

An arctic fox searches for food along the rocky beach of St. Paul Island in the Bering Sea. Photo courtesy of Chris Krenz.

COLOR: Alaska wolves vary in color from white to black, with various shades of gray and tan. The most common colors are gray and black. The wolves in Southeast Alaska are generally darker, and their pelage (fur) is courser than wolves found in interior Alaska. Wolves have a long, bushy tail and short ears similarly colored to the rest of their pelage.

SPEED: Wolves have long legs and are excellent runners. They can reach maximum speeds of 40 miles per hour (64 km/hr) or maintain a slower trot or lope for extended periods of time.

LONGEVITY: Wolves can live 12 years in the wild.

REPRODUCTION: Usually only one pair, the alpha pair or "top dogs," of a pack will mate. Wolves enlarge and often use an old fox den. The den is up to 10 feet (3 m) in length in well-drained soil. Breeding takes place February to March, and litters of two to 10 pups are born from early May to early June. By midsummer the pups are about three-quarters of adult size and are ready to travel with the pack. Pups reach full body size and weight in another year and may begin breeding at between two and five years of age. The alpha pair mates for life unless one of the pair dies, when the surviving wolf may take a new mate.

SOCIAL STRUCTURE: Wolves are pack animals and the alpha male and female are pack leaders. Packs usually consist of two to 12 members, but some winter packs have more than 30 members.

DISTRIBUTION: Wolves are found throughout mainland Alaska, many islands of Southeast Alaska, and Umiak Island in the Aleutians.

MOVEMENTS AND MIGRATION: A pack of wolves can maintain a territory of up to 3,000 square miles (7,770 km^2). Within that territory they may travel more than 20 miles (32.2 km) per day in search of prey. The size of the territory and the daily movements of the pack are dependent upon the prey abundance, habitat type, and time of the year. Some packs leave their regular territory to follow migrating herds or to move into areas with abundant prey, such as near salmon streams during spawning season.

A WOLF WAITS AT ITS DEN.
PHOTO COURTESY OF US FISH AND WILDLIFE SERVICE.

BLACK WOLF.
PHOTO COURTESY OF US FISH AND WILDLIFE SERVICE.

PACK OF WOLVES TRAVEL SINGLE FILE IN THE SNOW.
PHOTO COURTESY OF US FISH AND WILDLIFE SERVICE.

HABITAT: Wolves are successful in a variety of habitats from the coastal rain forests to the open tundra of the interior of Alaska. Accordingly, wolves can survive on a variety of prey.

PREY: Wolves are carnivores feeding mostly on large mammals such as moose, bison, elk, caribou, Dall sheep, mountain goat, black-tailed deer, seal and beaver. They also will take smaller prey such as voles, lemmings, squirrels, hares, birds and fish. Wolves have an amazing ability to go for extended periods with little food. A wolf pack needs to prey on large herbivores for sustenance and to maintain pack integrity.

PREDATORY CHARACTERISTICS: Wolves have keen senses and are very intelligent. They most often hunt as a team, which allows them to take prey much larger than they are. Large prey, such as moose, bison and elk, usually are surrounded and attacked from several sides at once. The killing usually involves attacks to the throat. Wolves have large canine teeth for securing and killing prey. Their massive jaw muscles and scissor-action molars allow them to slice through flesh and to break large moose bones so as to extract the marrow. Wolves of North America usually avoid human contact, whereas wolves of Eurasia are more aggressive towards people.

WOLF WAITING FOR PREY.
PHOTO COURTESY OF US FISH AND WILDLIFE SERVICE.

PREDATORS: Trapping by people, and intraspecific strife (among wolves) keep wolf numbers low. Wolf mortalities also result from disease, starvation and injuries acquired in capturing large prey. Bears will kill wolves.

CURRENT STATUS: Wolf populations are generally healthy in Alaska, but population densities vary depending upon prey abundance, disease and human activity. Population densities rarely exceed an average of one wolf per 25 square miles (65 km^2).

WOLF TRACKS.
PHOTO COURTESY OF US FISH AND WILDLIFE SERVICE.

ECOLOGY/CONSERVATION: Wolves are the top terrestrial predator in Alaska. They are considered a keystone species because they influence their prey community at a rate greater than expected from their biomass. This top-down control is somewhat mitigated by their taking weak animals when they can. When wolves hunt, which is most of the time that they are on the move, they detect their prey (visually or by smell) by chance encounter. The pack will follow the trail until the prey is encountered, then surround the prey to test its status, health and escape capabilities. Many healthy or fit prey animals will not flee, and instead will confront wolves. Attacking fit prey is dangerous, and the wolves often leave these animals and search for easier prey. Other fit prey may run and simply outdistance the predators. Hunting strategies used by wolves are most effective on sick, old and otherwise unhealthy prey. This is beneficial for the remainder of the herd in reducing the overgrazing of limited winter forage or in reducing the threat of disease spreading to other members of the herd.

In Alaska, moose and caribou populations can be reduced by severe winters and further reduced by wolf and bear predation; they also are reduced by human hunting activities. Continued hunting and predation may keep the prey numbers low in what is referred to as a "predator pit." When prey populations are reduced, wolves have few options for survival. Some predators can move to other areas. Wolves are very territorial, so this option usually is not available to them. Switching prey to snowshoe hare, sheep or some other prey is another option. A pack's decline may be postponed or avoided if the pack can sustain itself on buffer species such as beaver and salmon.

Declines in large ungulate populations often put wolves in direct conflict with people who also hunt and eat moose and caribou. Historically, predator-control programs have been common in Alaska with directed trapping, hunting and other predator-control measures used to reduce the wolf population.

WOLF.
PHOTO COURTESY OF US FISH AND WILDLIFE SERVICE.

Proponents of predator-control programs justify their strategy, stating that more stable and higher prey populations are best for all concerned, the moose, caribou, wolves and people. They argue that maintaining the ungulates at a high alternate stable state can be maintained, but this is not supported by the science. In Alaska, major population fluctuations are the rule and stable populations are not.

During the thousands of years that wolves have occupied Alaska they have had to cope with natural cycles of high and low prey abundances. These swings in prey density have exerted a positive evolutionary influence on predators and prey alike.

Only the strong survive — and they have. Because of these critical periods of reduced prey abundance, wolves, moose and caribou have evolved into what they are today. Strict resource management, which incorporates predator control may eliminate the positive effects of the natural cycles in the ecosystem.

Wolves inspire many feelings in people: awe, fear, love, and hate, among others. Fear and hate can be attributed to the lack of knowledge and understanding of wolves, a very intelligent and social species. Many people are in awe of the wolf's ability to survive in the wilderness while maintaining intact social packs. The call of the members of a wolf pack can be interpreted by some people as a call for the protection of wildness and wildlife. Most who hear the howls of wolves feel a sense of respect for these predators.

Black Bear

Black bears (*Ursus americanas*) are classified in the order Carnivora along with wolves, wolverine and lynx. However, bears are not carnivores; they are omnivores, eating both plants and animals. Some people erroneously believe that bears are closely related to pigs, partially because both species are omnivores. Female bears and pigs are called sows, male bears and pigs are called boars, and both species harbor the parasite that causes trichinosis. From an evolutionary standpoint, bears and pigs are not related. Pigs are classified in the order Artiodactyla, along with deer, goats, sheep and giraffes, because they have hooves, not paws and claws, like bears. Furthermore, a bear's teeth and leg bones, characteristics used to classify animals, are not like those found in pigs.

SIZE: Male black bears are larger than the females. The average Alaska male black bear is about 2½–3 feet (0.7–1.0 m) tall at the shoulder and five feet (1.5 m) long: add a few inches (8 cm) for the tail. Black bears may weigh 40 percent less in the spring after emerging from their winter dormancy, a period during which they do not eat. The average spring weight of a female black bear is 100–150 pounds (45–60 kg), and the average adult male spring weight is 170–200 pounds (77–90 kg). Spring males weighing nearly 400 pounds (180 kg) have been reported in Alaska. Fall male black bears can weigh over 500 pounds (227 kg).

COLOR: Black bears are usually black, but additional color variations include cinnamon, gray or blue. The gray or blue-colored black bear is often referred to as a "glacier bear." A white color phase occurs in British Columbia but has not been reported in Alaska. Any color sow can give birth to cinnamon, black or glacier-colored cubs. More than one color phase may occur in the same litter. Black bears often have a white patch on the front of their chests.

SPEED: Black bears can reach speeds of 30 miles per hour (48 km/hr).

LONGEVITY: The life span for black bear is approximately 20 years in the wild.

REPRODUCTION: During spring, boars will approach sows cautiously, moving in a non-threatening manner making sure their body language does not frighten the sow. Boars tend to be submissive and playful around the sows at this time of the year. Once the female becomes comfortable with the boar, mating may occur several times a day for several days. After this time they go their separate ways in search of food or other mates. Boars attempt to mate with as many sows as possible. The sow may mate with more than one boar.

Fertilized eggs stay in a suspended state until fall,

then implant in the uterus if the sow is healthy enough to support herself and the cubs through the winter. This is referred to as "delayed implantation." After implantation the embryos will develop for seven months. Cubs are born in the winter den when the female is in winter dormancy. She may give birth to one to four cubs, small, less than one pound (<0.5 kg), blind and nearly hairless. Cubs begin suckling right away and grow quickly.

Cubs weigh about five pounds (2.3 kg) when they emerge from the winter den, covered with a fine thick coat. The youngsters must learn how to locate food and avoid predators and other dangers. They remain with the sow for that summer, fall and another winter, taking survival lessons from their mother. When the sow breeds again the following summer, the cubs are usually on their own. In the more northern regions of Alaska, where the conditions are more difficult, cubs stay an additional year with the sow, and she may only breed every three years. Black bears become sexually mature when they are between three and six years old.

SOCIAL STRUCTURE: Black bears are usually solitary except for sows with cubs and during mating season in June and July. Sows with cubs are cautious around boars and avoid them. Boars kill cubs if given the chance. Since these sows would then soon come into heat again, it allows the boar an evolutionary advantage for him to pass on his genes to the next generation.

Bears may gather at abundant food sources such as salmon spawning streams. A hierarchy is established at the site, with the dominant animals controlling the best fishing spots. The most dominant animals are the large boars followed by sows with older cubs then sows with younger cubs and then other younger bears. Body language is used to establish and maintain rank, which reduces the number of physical confrontations.

DISTRIBUTION: Black bears are found in most of Alaska's forested areas. They are not found on the Seward Peninsula, north of the Brooks Range, or on many of the large islands of the Gulf of Alaska (including Kodiak, Montague, Hinchinbrook, Admiralty, Baranof, Chichagof and Kruzof) where brown bears are abundant.

THIS LEAN (SPRING) BLACK BEAR IS LOOKING FOR FOOD. *PHOTO COURTESY OF JOHN GOMES.*

THIS SOW IS LYING AMONG ONE OF THE BLACK BEAR'S FAVORITE EARLY SUMMER FOODS—DANDELIONS. *PHOTO BY BRUCE WRIGHT.*

THIS SOW BLACK BEAR AND HER CUBS WERE SEEN ALONG THE ALASKA HIGHWAY. *PHOTO BY BRUCE WRIGHT.*

SEQUENCE OF PHOTOS OF A BLACK BEAR RUNNING DOWN A HILL. PHOTOS COURTESY OF JOHN GOMES.

MOVEMENTS AND MIGRATION: Bears leave their winter dens in the mountains in spring, and by July most black bears are feeding on the lush vegetation found in meadows and avalanche chutes. Bears learn from their mothers and from their own experiences where and when to find food. This drives their annual wanderings. Bears depend upon the entire landscape for survival. The first emerging plants in spring are low in the valleys and along coastal shorelines. As spring moves up the hills and mountains, bears move up with it to eat nutritious emerging plants. By late summer many black bears are again at higher-elevation meadows feeding on the lush plants just released from winter's grip.

Black bears will move to where salmon and other foods are available. This brings them back down the mountain to salmon streams and berry patches. Bears often use the same corridors, or paths, as they move through the forest. These well-used trails have worn footsteps, sometimes in solid rock, where many generations of bears have passed.

By late fall most black bears are again moving back up the mountain, feeding on the ripening berries. As snow begins to build up in the mountains in October, black bears start locating and preparing their winter dens. Most black bears will stay in their dens all winter, but some bears will come out for a brief trip and then return to stay in their dens until spring.

HABITAT: Black bears usually are associated with forested habitat; however, bears are creatures of landscape in that they use a wide variety of habitats including streams, forested valleys, beaches, meadows, muskegs, tundra and mountains. They are rarely seen on the open tundra, probably because there are no trees to climb to escape predation by grizzly bears.

PREY: Bears need to eat enough in a few months to provide energy to survive the long winter. Black bears are mostly vegetarians and may remind the human observer of cows as they are seen grazing on sedges and fern fiddleheads. However, black bears also will eat winterkilled animals and sometimes will kill and eat newborn moose calves and deer fawns. They have been seen digging and eating clams.

In August salmon are on the menu for most black

bears that can access a salmon spawning stream. Salmon can be a boon for bears by supplying them with necessary nutrients and calories. The eggs, brains and skin of salmon are packed with calories, and it is calories that are needed to get black bears through the winter. Bears usually eat the entire fish early in the season but, as salmon spawning season progresses, bears become picky and may eat only the brains and eggs, discarding the muscle tissue completely. Bears often discard male salmon in favor of egg-laden females.

PREDATORY CHARACTERISTICS: Black bears occasionally prey on young moose, caribou, deer and other ungulates. They may happen upon these younger animals, but some black bears learn to hunt them. The bears' sharp claws and long canine teeth are used to secure and kill their prey. Black bears sometimes stalk and kill people, although this is rare.

PREDATORS: Wolves or brown bears may kill black bears, especially those that are younger and smaller. Sometimes male black bears will kill bear cubs, possibly for food. People hunt black bears for food and for their hides.

CURRENT STATUS: Black bear populations are healthy in Alaska, although bears often are driven away from human populated areas. In other urban areas, bears are attracted to human waste and garbage.

ECOLOGY/CONSERVATION: The next time you are in the Alaska woods and come across a large pile of bear feces, grab a stick and pick it apart. To some this may seem odd and disgusting, but to those interested in nature and in bears, poking and prodding a large pile of bear scat can be revealing and interesting. The salmon bones are easy to pick out, and you even may be able to determine if the bear has been eating chum or pink salmon by the size of the bones. Bones from mice, rabbits and other mammals can indicate a wider feed source. Most green vegetation is black after being digested, but the berries can be very interesting. Blueberries can turn the feces dark purple to black, but many of the berries can pass through the bear unscathed. When bears eat berries they usually eat some of the leaves. These can be seen in the mix.

BEAR SCAT WITH BERRIES.
PHOTO BY BRUCE WRIGHT.

Black and brown bears may eat high-bush cranberries or devil's club berries. These red berries actually make a pile of bear scat rather attractive. Older piles of bear scat, from previous years, sprout mushrooms or seedlings of the berries the bear ate. That is how many berry bushes get their start in life.

All bears have good senses of sight and hearing. Their sense of smell is excellent. Alert bears can detect, with their eyesight alone, a moving human form from over a mile (1.6 km) away, and move off to avoid contact. Use this information to avoid dangerous encounters with bears. If you spend time in bear country you should:

- Look for signs of bears and make plenty of noise.
- Avoid surprising bears at a close range.
- Avoid crowding bears; respect their "personal space."
- Avoid attracting bears through improper handling of food or garbage.
- If confronted by a bear, plan ahead, stay calm, identify yourself by yelling (e.g., "Hey, bear! Scram!") and don't run.

In most cases bears are not a threat, but they do deserve your respect. More information about what to do and how to operate in bear country can be obtained from the Alaska Department of Fish and Game www.wildlife.alaska.gov/aawildlife/harmony.cfm.

Black bears usually avoid contact with humans. However, once black bears become habituated to humans and to their food they may become especially dangerous. These bears lose their fear of humans and associate humans with food. Each year many bears are killed in Alaska's towns and cities because they become accustomed to feeding on garbage, and are considered unpredictable and dangerous to humans. Providing food for bears usually results in the bear being in conflict with humans. Capturing habituated or "garbage bears" and relocating them often does not work. Some bears have returned over 100 miles (161 km), crossing major rivers, while other bears find another community and, in both cases, they resume their bad habits. Protect bears by eliminating garbage and other available food items from camps and residences.

Hibernation is something arctic ground squirrels do, when their body temperatures drop to just above freezing and the heart rate is a small fraction of normal. Black and brown bears go into winter dormancy, their body temperatures drop a few degrees, their metabolic rate is reduced, and they sleep for long periods. In the more southern ranges black bears occasionally will emerge from their dens during winter. In the northern part of their range bears may be dormant for as long as seven or eight months. Females with cubs usually emerge from winter dormancy later than boars or sows without cubs.

Black bear dens in Alaska are often in rock cavities or caves, fallen hollow trees, or dug into dirt or clay banks. Most dens are in the mountains, but some bears use dens at lower elevations. Black bears may prefer to hibernate in the mountains and on north-facing slopes to avoid melt water flooding the den when warm, wet winter storms come in off the ocean. During winter dormancy bears don't urinate, defecate or eat. They must survive off their fat reserves.

Salmon return from the sea, enter their natal rivers and streams, spawn and die. Were it not for the predators, such as bears, that eat them many of the nutrients found in the dead salmon would be washed back to sea. Black bears and other predators are important for moving the nutrients into the watershed. Their scat enhances the growth of trees and vegetation along the length of the streams where salmon spawn. The entire forest ecosystem is dependent upon bears and other predators for fertilizing the watersheds with nutrients from the salmon. Even the salmon benefit because the fertilized watershed will produce more stream-shading vegetation and more insects used as food by young salmon. The productivity of salmon is enhanced in systems with bears.

Black bears are important food for humans in some parts of Alaska. Spring bears are lean and their flesh is usually very palatable. Late summer and fall bear meat may be less tasty, especially when they have been feeding on salmon. Sometimes fall bears that have fattened up on berries are harvested just for their fat, which is rendered and used in cooking. Good-quality bear fat looks just like white pig lard and is used for baking pies and donuts. Black bears can harbor the parasite that causes trichinosis. To prevent infection from this disease, wear protective gloves when handling bear meat, and cook the meat thoroughly before eating.

Brown Bear

Brown bears (*Ursus arctos*) are also called "grizzlies" or "grizzly bears." The term "grizzly" usually is used for brown bears that occur more than 20 miles (32 km) from the coast. They are the same species, but grizzly bears usually have less access to salmon and high-quality food. Grizzly refers to the silvery or "grizzled" highlights in the pelage of some brown bears.

SIZE: Alaska's brown bears can be very large. Boars can stand over nine feet (3 m) tall and weigh in at 1,400 pounds (635 kg). Brown bears may weigh 40 percent less in the spring after emerging from their winter dormancy, a period during which they do not eat. The average spring weight of a boar brown bear is 500–900 pounds (227–408 kg). The average adult sow brown bear's weight is about half that of the male's.

COLOR: Brown bear colors range from almost black to

brown and even light blond. The lighter-colored brown bears usually are found in interior Alaska while the dark brown animals are more often found along coastal regions.

SPEED: Brown bears can reach speeds of 35 miles per hour (56 km/hr).

LONGEVITY: Brown bears are reported to live at least 34 years in the wild; however, brown bears are considered to be old in their early to mid-twenties.

REPRODUCTION: Brown bears are usually solitary, except for sows with cubs and during mating season from May to July. Sows are cautious around boars and usually avoid them. Boars cautiously approach sows in the spring, moving in a non-threatening manner. Boars often will be submissive and playful around the sows during this time of the year. The boar attempts to mate with as many sows as possible. The sow may mate with more than one boar.

Fertilized eggs will stay in a suspended state until fall, then implant in the uterus if the sow is healthy enough to support herself and the cubs through the winter. This is referred to as delayed implantation. After implantation the embryos will begin to develop. The cubs are born in the winter den in January or February when the female is in winter dormancy. She may give birth to one to four cubs, but a litter of two is most common. When born, the cubs are small, less than one pound (<0.5 kg), blind and nearly hairless. They begin suckling right away.

When they emerge from the winter den, cubs weigh about five pounds (2.3 kg) and are covered with a thick coat of fine hair. Brown bears must learn how to locate food and avoid predators and other dangers, lessons they learn from their mother. They remain with the sow for two summers. In the more northern parts of Alaska, cubs stay with the sow for as long as five years. The sow will not mate until after the cubs have left her. She may produce another litter the following summer, but where food is scarce, she may not breed again for several years.

SOCIAL STRUCTURE: Brown bears may gather at abun-

A BROWN BEAR SEARCHING FOR AND CATCHING A SALMON IN AN ALASKA STREAM.
PHOTOS COURTESY OF JOHN HARVEY.

dant food sources such as salmon spawning streams. A hierarchy is established at the site with the dominant animals controlling the best fishing spots. The most dominant animals are large boars followed by sows with older cubs then sows with younger cubs, and then other younger bears. Body language is used to establish and maintain rank, which reduces the number of physical confrontations. Fights do occur, and some bears have large scars from these encounters.

A BROWN BEAR IN THE FALL, THE TIME OF THE YEAR THEY ARE HEAVY WITH FAT.
PHOTO COURTESY OF US FISH AND WILDLIFE SERVICE.

A SOW AND HER CUB WALKING IN AN ALASKA STREAM.
PHOTO COURTESY OF JOHN HARVEY.

THIS BROWN BEAR IS FEASTING ON SEDGES GROWING ALONG THE BEACH. PHOTO BY BRUCE WRIGHT.

DISTRIBUTION: Brown bears occur throughout Alaska, except on the islands south of Frederick Sound in Southeast Alaska, the islands west of Unimak in the Aleutian Chain, and the islands of the Bering Sea. Kodiak, Montague, Hinchinbrook, Admiralty, Baranof and Chichagof islands all have high densities of brown bears. Admiralty Island, the third largest island in Southeast Alaska's Alexander Archipelago, has about one brown bear per square mile (2.6 km^2). The local Tlingit Indians call this island Kootznoowoo which means "fortress of the bears."

MOVEMENTS AND MIGRATION: Depending upon the winter and spring conditions and the location, brown bears emerge from their dens from April to June. Boars usually leave their dens first and move to lower elevations in search of food. During spring, bears often feed on sedges, a lush, grass-looking plant growing in wet areas. Sows with older cubs emerge next and move down the mountain, while sows with new cubs emerge last.

Bears learn from their mothers, and from their own later experiences, where and when to find adequate food. This drives their annual wanderings. Bears depend upon the entire landscape for survival. They detect the first emerging plants in the spring, low in the valleys and along coastal shorelines. As spring moves up the hills and mountains, bears move up to eat the more nutritious emerging plants. By late summer many brown bears may again be at higher-elevation meadows feeding on the lush plants just released from winter's grip.

Brown bears will move to where salmon and other foods are available. This brings them back down the

BROWN BEAR TRACKS ON AN ALASKA BEACH.
PHOTO BY BRUCE WRIGHT.

A BROWN BEAR EMERGING FROM THE FOREST INTO A MEADOW.
PHOTO COURTESY OF JOHN GOMES.

mountain to salmon streams and berry patches. Bears often use the same corridors, or paths, as they move through the forests. These well-used trails have worn footsteps where many generations of bears have passed. These trails sometimes can be worn in solid rock. By late fall most brown bears again are moving back up the mountain, feeding on the ripening berries. The snows begin to build up in the mountains in October. This is when brown bears start locating and preparing their winter dens.

HABITAT: Brown bears usually are associated with open habitat; however, bears are creatures of landscape in that they use a variety of habitats including streams, forested valleys, beaches, meadows, muskegs, tundra and mountains.

PREY: Brown bears are opportunistic and will eat winterkilled animals and sometimes kill and eat moose, caribou and deer. They have been seen eating clams and mussels.

Salmon is on the menu for most Alaskan brown bears in late summer and fall. The eggs, brains and skin of salmon are packed with fats and calories, and it is calories that are needed to get bears through the winter. Bears usually eat the entire fish early in the

BROWN BEAR IN A MEADOW.
PHOTO BY BRUCE WRIGHT.

season, but as salmon spawning season progresses, bears become picky and may eat only the brains and eggs, discarding male salmon completely.

PREDATORY CHARACTERISTICS: Brown bears occasionally prey on ungulates (moose, caribou, deer, etc.). Some brown bears learn to hunt specifically for the youngsters during the spring calving season.

THIS BROWN BEAR HAS FOUND A SAFE PLACE TO EAT ITS CATCH.
PHOTO COURTESY OF JOHN HARVEY.

PINK SALMON WITH THE BRAINS EATEN BY A BEAR.
PHOTO BY BRUCE WRIGHT.

THIS BEAR HAS CAUGHT A PINK SALMON.
PHOTO COURTESY OF SCOTT GENDE.

Brown bears' long claws and canine teeth are used to secure and kill their prey. Brown bears use their long claws for digging roots, rodents and other prey from their dens. Brown bears sometimes kill and eat people, although this is rare.

PREDATORS: Wolves or other brown bears may kill brown bears, especially younger, smaller animals. People hunt brown bears for their hides and as trophies.

CURRENT STATUS: Brown bear populations are healthy in most areas of Alaska. However, near some communities and heavily populated areas (Anchorage, Juneau and on the Kenai Peninsula) brown bear populations nearly have been eliminated or reduced to critical levels because of their negative interaction with people.

Bear movements and population densities vary depending on food availability and hunting pressure. In areas of low productivity, such as on Alaska's North Slope, bear densities are as low as one bear per 300 square miles (777 km^2). In areas with easily obtainable food, such as Kodiak and Admiralty Islands, densities are as high as one bear per square mile (2.6 km^2).

ECOLOGY/CONSERVATION: Brown bears have a very low reproductive rate; interactions with people easily can cause a reduction in brown bear populations. Some bear encounters with humans result in the death of the bear; more encounters certainly correlate to more bear deaths. One of the most common places where people encounter bears is on roads in remote areas. Logging roads are built for the removal of timber, and often are left intact after logging is completed, leaving convenient paths for bears and people alike. A case study of the brown bears on northern Chichagof Island of Southeast Alaska revealed that a large number of bears are killed in the protection of human life and property, and that the majority of the bear encounters and deaths are on the old logging roads in the area. One of the best ways to protect wildlife, not just bears, is to reduce human access to wilderness areas. The best way to control and reduce human access is to close or eliminate roads into these areas.

Some brown bears become habituated to humans and to their food. These bears may become especially dangerous because they learn to associate people with food, and they lose their fear of humans. Bears are killed in Alaska's towns and cities every year because they have been allowed to eat human food and, as a result, may become dangerous. Providing food for bears usually results in the bear's death. Protect bears by eliminating available garbage and other food items from camps and residences.

A BROWN BEAR ATTEMPTING TO CATCH SALMON IN AN ALASKA STREAM. *PHOTO COURTESY OF JOHN GOMES.*

THIS PHOTOGRAPHER IS TAKING PHOTOS OF A BROWN BEAR USING A TELEPHOTO LENS. *PHOTO BY BRUCE WRIGHT.*

THE FISHERMAN IS UNAWARE OF THE BEAR THAT IS ALSO INTERESTED IN CATCHING SOME FISH. *PHOTO COURTESY OF JOHN GOMES.*

THIS LARGE, MALE BROWN BEAR WAS TRANQUILIZED SO SCIENTIFIC DATA COULD BE COLLECTED. THE BEAR WAS TAGGED AND SAFELY RELEASED. PHOTO COURTESY OF SCOTT GENDE.

There are many notable differences between brown and black bears. Black bears have a straight facial profile and no large shoulder hump. Brown bears have a large shoulder hump, muscles used for digging. If you get close enough you may notice that black bears have claws that are sharply curved and seldom over 1½ inches (3.8 cm) in length, while brown bears' claws are long, as long as 4 inches (10 cm), and they are straight. Black bears' claws are retractable like a cat's, while brown bears' are not. If you get really close you may note that the black bear's upper rear molar is never more than 1¼ inch (3.2 cm) long, while an adult brown bear's upper rear molar is never less than 1¼ inches (3.2 cm) long.

All bears have good senses of sight and hearing. Their sense of smell is excellent. Alert bears can detect, with their eyesight alone, a moving human form from over a mile (1.6 km) away. Once a human is detected, the bear usually moves off to avoid contact. Use this information to avoid dangerous encounters with bears. If you spend time in bear country you should:

- Look for signs of bears and make plenty of noise.
- Avoid surprising bears at close proximity.
- Avoid crowding bears; respect their "personal space."

- Avoid attracting bears through improper handling of food or garbage.
- If confronted by a bear, plan ahead, stay calm, identify yourself as a human by yelling (e.g., "Hey, bear! Scram!") and don't run.

In most cases bears are not a threat, but they do deserve your respect. More information about what to do and how to operate in bear country can be obtained from the Alaska Department of Fish and Game, www.wildlife.alaska.gov/aawildlife/harmony.cfm.

Brown bears are rarely used for human food. When asked if he liked eating the meat of a large brown bear he shot, the hunter said, "The meat tastes pretty bad, but you sure get a lot of it!" Bears can harbor the parasite that causes trichinosis. To prevent infection from this disease, wear protective gloves when handling bear meat, and cook the meat thoroughly before eating.

Brown bears have a low reproductive rate. Because of this, bears easily can be over-harvested. Hunters should avoid harvesting females.

Lynx

Lynx (*Lynx Canadensis*), or "link" in parts of Alaska and the Yukon Territory, is the only common wild cat in Alaska. Lynx are shy and rarely seen, so most people believe them to be scarce.

SIZE: Lynx have long legs, large furry feet and long tufts on the tip of each ear. The large feet act as snowshoes, allowing the cat to move easily over soft, deep snow. Males can weigh up to 40 pounds (18 kg) and are larger than the females (18–30 pounds or 8–14 kg).

COLOR: Lynx have black-tipped short tails, similar to bobcats. The lynx tail tip is completely black all around, while bobcat tails show black bars with a white tip when viewed from above, and white underneath. The soft fur of the lynx is very luxuriant, buff gray and tan with spots. The underbelly is lighter than the back and sides.

LYNX.
PHOTO COURTESY OF JOHN GOMES AND THE ALASKA ZOO.

SPEED: Lynx can reach speeds of 30 miles per hour (48 km/hr).

LONGEVITY: Lynx can live 15 years in the wild.

REPRODUCTION: Mating usually begins in March and runs into April. The gestation period is 63 days. The two to six kittens are born in a natural shelter such as a group of fallen trees. The size of the litter and their survival is correlated to food availability, especially snowshoe hares. Their eyes open after one month and they are weaned at two to three months. Kittens remain with

LYNX.
PHOTO COURTESY OF JOHN GOMES AND THE ALASKA ZOO.

LYNX.
PHOTO COURTESY OF JOHN GOMES AND THE ALASKA ZOO.

their mother through the winter to learn how to hunt and survive in the wild. If food is abundant, all the kittens are likely to survive. The family members go their separate ways during the following breeding season. Lynx become sexually mature after one year of age.

SOCIAL STRUCTURE: Lynx are usually solitary, except that kittens will often stay with their mother for the first winter.

DISTRIBUTION: Lynx occur throughout most of Alaska and Canada, primarily in boreal forests. The only places where they have not been sighted in Alaska are on offshore islands from the Bering Sea to Southeast Alaska.

MOVEMENTS AND MIGRATION: A lynx's home range is 5–100 square miles (13–260 km^2), depending upon terrain and prey availability. Males usually have larger home ranges than females. Within this home range they normally travel one to five miles (1.6–8 km) per day. Lynx do not migrate but may wander long distances to locate prey. During years when hare populations crash they will show up in areas where they are not usually seen, sometimes crossing the coastal mountains looking for food.

HABITAT: Lynx usually use forested or brushy areas, wherever they can find snowshoe hares.

PREY: Snowshoe hares are the primary prey for lynx, but they will prey upon grouse, ptarmigan, squirrels and microtine rodents (meadow mice or voles). During periods when hare abundance is low, lynx will prey on caribou, sheep and foxes.

PREDATORY CHARACTERISTICS: Lynx are heavily reliant on their extraordinary eyesight for hunting. They hunt mostly at night, but they also are successful daytime hunters. Hare can detect sounds and ground vibrations that help them to avoid predators. Lynx stalk quietly through the forest searching for prey. Another hunting technique used by lynx is to wait along the trail and ambush hares as they move about searching for food.

If a chase ensues, the broad feet of the lynx help it to remain on top of the snow. The capture must occur quickly or the hare is likely to escape. Some chases reach speeds of 30 miles per hour (48 km/hr). When caught by a lynx the hare often gives a loud squeal, but is quickly silenced with a strong bite to the neck. The hare is consumed on the spot or taken to a more secure area. An adult lynx needs to eat one hare every day or two.

LYNX.
PHOTO COURTESY OF JOHN GOMES AND THE ALASKA ZOO.

PREDATORS: Bears, wolves and other lynx may prey upon lynx. Humans trap them for their fur.

CURRENT STATUS: The Alaska lynx population is healthy. The main human threats to lynx are habitat destruction and trapping. Large fluctuations in regional lynx populations are common and natural, because they are dependent upon the fluctuating snowshoe hare population in the area.

ECOLOGY/CONSERVATION: Snowshoe hare do best in habitat that contains a mixture of vegetation types, usually an area that has been recently burned and which has early successional growth. The snowshoe hares find this new growth more available (low to the ground) and more nutritious. When the hare cycle is high and the habitat is good, this prey base supports a healthy predator population, including lynx. Lynx populations will increase dramatically when the hare population peaks. One or two years after the hare population crashes, lynx and other predator populations also will crash. Snowshoe hare populations are not only controlled by habitat quality, but by other predators such as birds of prey and mammals that hunt hare. The lynx and hare population cycles, which occur every eight to ten years, have been documented in trapping records for centuries and have been studied carefully by scientists.

MINK

The mink (*Mustela vison*) is a member of the weasel family, Mustelidae, a group of renowned predators capable of taking prey many times their size. Other members of the Mustelidae family include marten, otter, weasel and wolverine.

THIS MINK WAS SEARCHING FOR PREY ON A BEACH IN SOUTHEAST ALASKA. PHOTO BY BRUCE WRIGHT.

MINK TRACKS. PHOTO BY BRUCE WRIGHT.

SIZE: Mink have a long slender body, short legs, and a six- to nine-inch (15–23 cm) tail. Males are about two feet (60 cm) long, longer than the females by a few inches (7–8 cm). Males weigh two to four pounds (0.9–1.8 kg) and females one to three pounds (0.5–1.4 kg).

COLOR: Mink have a beautiful, dark brown fur coat with a small white patch on the chin, throat and chest area.

SPEED: Mink use stealth rather than speed to capture prey and avoid predation. Like marten, mink can reach speeds of 25 miles per hour (40 km/hr), but only for short distances. Mink are semi-aquatic and have webbing between their toes to aid in swimming. Mink are good swimmers, but also must use cunning and stealth in the water to be successful predators.

LONGEVITY: Mink can live 10 years in the wild.

REPRODUCTION: Breeding season for mink is from the late winter to early spring. They mate several times with multiple mates. Fights and disputes sometimes occur over mating rights. Mink have delayed implantation, and the egg may begin development several months after mating. Once the egg implants, the gestation period is about 30 days. The female selects a den where she can access food easily, usually near water,

MINK. PHOTO BY BRUCE WRIGHT.

MINK. PHOTO BY BRUCE WRIGHT.

THESE ARE MINK PELTS; THE MINK WERE CAUGHT AND PREPARED BY A FUR TRAPPER.
PHOTO COURTESY OF THE US FISH AND WILDLIFE SERVICE.

and which is safe from predators. The den can be dug into the bank of a river or lake; or the old dens of other mammals, such as river otters, may be used. Mink use dried grass, leaves, moss and fur to line the area of the den where the young are born.

A litter of one to 10 young are born in late spring (April to May). The number and survival of young depends upon the food quality and abundance. At birth the young weigh about 1/3 ounce (9 grams), are pink, blind and covered with a thin, light-colored fur. Their mother's milk is high in fat and the young grow quickly. At about three to four weeks they open their eyes, and at about six weeks they are weaned. The young remain with their mother for their first summer and fall. Mink will breed during their first winter or spring.

SOCIAL STRUCTURE: Mink are primarily solitary animals, and males will attempt to exclude other males from their territories. Much growling and fighting may occur during a territorial dispute and during the mating season. Territorial boundaries are marked using secretions from their anal glands.

DISTRIBUTION: Mink are found in every part of the state, with the exceptions of the Aleutian Islands, Kodiak Island, the offshore islands of the Bering Sea, and most of the North Slope.

MOVEMENTS AND MIGRATION: The home range of a male mink is about two to three square miles (5.2–7.8 km^2) and usually is along shorelines of the ocean, rivers or lakes, although both sexes commonly travel considerable distances from the water. Male mink have a larger home range than do females.

HABITAT: Mink are common along the coast and shorelines of the ocean, rivers and lakes. They thrive near water where they find prey.

PREY: Mink are strictly meat-eaters (carnivorous) and will kill and eat fish, birds as large as geese, bird eggs, sea urchins, crabs, and small mammals as large as snowshoe hare. Increases in hare and vole populations may draw mink inland some distance from the water.

PREDATORY CHARACTERISTICS: Mink hunt primarily at night, but in coastline areas they sometimes are seen during the day, working the intertidal area in search of sea urchins and other prey at the low-tide line, or scurrying around the rocks of a seabird colony. They are skilled swimmers and can swim underwater for 100 feet (30 m) at a depth of five feet (1.5 m) in search of prey. Mink are skilled climbers also, but not as arboreal as marten.

PREDATORS: In Alaska, humans trap mink for their fur. Additionally, bald eagles, owls, hawks, wolves, foxes, lynx and river otters may prey upon mink. However, mink will inhabit river otter dens, even when otters are present. The relationship between these two species requires further investigation.

CURRENT STATUS: Mink populations in Alaska are thriving. Trapping pressure may cause localized depletions. Establishment of sound trapping regulations can prevent these depletions.

ECOLOGY/CONSERVATION: During the fur-trapping heydays of the early 1900s, mink were introduced to several islands in Alaska, including islands in Prince William Sound. Subsequent predation by mink has reduced several seabird populations. A program to remove introduced fox from some islands has been successful in restoring populations of some bird species, including the endangered Aleutian Canada goose. A similar program should be considered for removing introduced mink.

MARTEN

The pine marten (*Martes Americana*), or American sable, is known for the sable coats made from their luxurious fur. Marten are in the family Mustelidae, a group of carnivores that includes such fearless predators as the mink, otter, weasel, wolverine and sea otter.

SIZE: Marten are 19–25 inches (48–64 cm) long, plus their tail of six to nine inches (15–23 cm). Males are about 25 percent larger than the females. Males weigh up to two pounds (1 kg).

COLOR: The marten's fur is soft and dense with shades from light blond to very dark brown. The legs and feet are usually darker brown or black. The throat patch varies from a pale buff to a deep orange, and is a striking contrast to the animal's body color. The ears are rounded and erect. Marten have the standard weasel look with a long snout and a long thin body. The large furry feet are good for walking on the snow, and the sharp claws make marten excellent climbers. The front claws are retractable like cats' claws.

SPEED: Marten use stealth rather than speed to capture their prey and to avoid predation, but they can reach speeds of 25 miles per hour (40 km/hr) for short distances.

LONGEVITY: Marten have been reported to live 17 years in captivity, but they rarely live 14 years in the wild.

REPRODUCTION: Male and female marten tend to have several mates. Females may mate at 15 months of age. The fertilized eggs do not implant in the uterus for about six months. This is delayed implantation. The litter contains two to five young and each is about one ounce (28 grams) at birth. They are born during April or May in a den in a hollow tree or in a pile of rocks.

MARTEN.
PHOTO COURTESY OF THE US FISH AND WILDLIFE SERVICE.

The den's nesting area is lined with dry grass, leaves and moss.

When born, the young are covered with a fine yellowish hair, their eyes are closed, and they are completely dependent upon their mother. Their eyes open when they are about five to seven weeks old, and by early fall they can forage by themselves. Marten reach their full length by three months and their full weight after one year.

SOCIAL STRUCTURE: Marten are solitary except for reproduction and when they gather at abundant sources of food such as salmon spawning streams or at large supplies of carrion, such as a moose carcass. Marten maintain a territory within their home range by scent marking and by forcefully excluding other marten. They scent-mark by rubbing their scent glands on rocks, stumps and trees. Rivers and shorelines may serve as natural home-range borderlines. The marten that cannot defend a territory are transients and may travel over 20 miles (32 km) per day in search of available habitat.

DISTRIBUTION: Marten are found throughout the forested areas of mainland Alaska and on many of Alaska's forested islands, where they were introduced by people.

MOVEMENTS AND MIGRATION: A marten's home range size, 1–15 square miles (1.6–24 km^2), is determined by the marten population density, type of habitat and prey density. When either the prey or marten density is high, the home range likely will be small. Males have larger home ranges than females. During their first year both males and females tend to wander, looking for available habitat.

Long-distance movements of many marten have been observed by trappers in Alaska, as evidenced by the sudden appearance of many marten on their traplines where previously there were few or none during that season.

HABITAT: Marten prefer to use mature coniferous forests where they seek cover, dens and food. They also use open meadows, old burned areas and grasslands where they locate adequate prey.

PREY: Marten primary prey is voles, a group of rodents whose populations can reach high densities. A marten needs to consume three voles each day, or an equivalent amount of prey. Marten have little body fat or energy reserves, and depend upon various strategies to stay fed, including caching food for later consumption

MARTEN.
PHOTO COURTESY OF JOHN GOSSE.

MARTEN.
PHOTO COURTESY OF JOHN GOSSE.

MARTEN. *PHOTO COURTESY OF JOHN GOSSE.*

THIS MARTEN WAS CAPTURED AND FITTED WITH A SATELLITE COLLAR THAT ALLOWED THE RESEARCHERS TO MONITOR THE ANIMAL'S MOVEMENTS. *PHOTO COURTESY OF JOHN GOSSE.*

or staying with large prey items, such as snowshoe hare, until the prey is consumed. They also readily will feed on winterkilled animals and animals killed by people and larger predators, such as wolves. Occasionally marten eat birds and their eggs. Marten also eat spawned-out and dead salmon, as well as berries and fruit. Trappers are well aware of this species' sweet tooth and will use sweet bait, such as strawberry jam. This helps the trapper avoid catching mink, a species that is not attracted to sweet bait, but can occur in the same habitat as marten.

Some marten steal food from people; persistent marten can harass hunters by trying to eat game meat hanging in camp. Marten sometimes enter remote cabins or use other human habitations in their search for food, but they also can help reduce problem mice around a cabin.

PREDATORY CHARACTERISTICS: Marten are voracious and opportunistic predators. One of their most successful hunting techniques is similar to that used by house cats. The marten will locate an area of high vole density and wait beside an active runway or hole until the victim moves into the open. The marten quickly pounces and dispatches the vole with its strong bite and sharp canine teeth.

PREDATORS: During peak snowshoe hare periods, the number of hare predators increases dramatically, especially lynx and great horned owls. Marten need to be especially cautious during these periods or they will be caught and eaten by some other Alaskan predator. Great horned owls, lynx, foxes and wolves prey upon marten, and people trap marten for their fur.

CURRENT STATUS: Marten populations are generally healthy in most areas of Alaska. However, near some communities and other populated areas, marten populations have been reduced by trapping.

ECOLOGY/CONSERVATION: A trapper can manage a marten population so as not to "trap out" an area. Marten are naturally curious and nearly always hungry, which makes them easy to catch. Marten have a home range of one to 15 square miles (1.6–24 km^2). Males have a larger home range than females, and juveniles tend to range over greater distances during their first year in search of their own territory. This wandering makes juveniles more likely to be captured by predators and trappers. Once the trapper begins catching adults, and especially adult females, the reproductive stock is being removed and the trapper should cease trapping in that area. Trappers can manage the marten populations actively and successfully by enumerating and monitoring the ratios of males to females and of juveniles to adults harvested. In a healthy population more males than females should be caught, and at least three juveniles should be caught per adult female. Juvenile marten are distinguished from the adults by slightly smaller ears, less jaw muscle mass (seen as a gap on top of the skull where the muscles do not meet in juveniles), and the juvenile males have a much smaller (approximately 25 percent) penis bone, or baculum.

Marten were introduced on several islands in Alaska to promote the fur industry. They should be considered an invasive species to the areas into which they have been introduced. Changes in the ecology have certainly taken place on many of Alaska's islands since the introduction of marten. In 1985 people introduced red squirrels to Admiralty Islands to serve as food for marten. By 1995 the squirrels had spread throughout the 35-mile by 65-mile (56 x 105 km) island. Ironically, marten rarely kill and eat red squirrels.

Humans can degrade marten habitat through development projects and clear-cut logging. New roads created in marten habitat, often associated with logging operations, allow human access into a new area, which allows trappers access to pristine marten habitat. Many species suffer adverse consequences when they come into contact with humans. Marten easily can be overexploited if people have access to them. Some of the best ways to protect predators and other wildlife is to limit human access, and the best ways to limit human access are to prevent building new roads, to restrict access on existing roads, or to remove the roads.

WOLVERINE

Wolverines (*Gulo gulo*) are the largest member of the land-dwelling Mustelidae family, which also includes weasel, mink and marten. In Alaska the wolverine is sometimes called the "devil bear."

SIZE: Male wolverines weigh 20–45 pounds (9.1–20 kg) and females are slightly smaller at 15–30 pounds (6.8–13.6 kg). They stand 12–16 inches (30–41 cm) tall at the shoulder.

COLOR: Wolverine fur is dense with long guard hairs. The fur is dark brown, almost black, with a horseshoe-shaped blond band running from shoulder to tail. Variations in color may exclude the blond patch on the back, or the entire animal may be blond. Most wolverines have a blond-to-orange throat patch. They have a long, bushy tail. Wolverine claws, which are sharp and curved, are ivory white, contrasting with their dark fur.

SPEED: Wolverines have relatively short, stout legs under a rounded, robust body. These features give them a characteristic hopping shuffle or gait, which usually does not exceed 10 miles per hour (16.1 km/hr). However, they are very persistent and regularly travel great distances in search of food.

LONGEVITY: Wolverines can live 13 years in the wild.

REPRODUCTION: Wolverines are mature at two years of age. They breed from May to August, often with more than one mate. The embryos do not implant in the uterus until early winter. This delayed implantation occurs only if the female is healthy. The female selects a den in the rocks, usually in the mountains. In most areas of Alaska the den site is covered with a thick blanket of snow. Some dens have snow tunnels that extend for over 50 yards (46 m). One to four kits are born between January and April. The female wolverine cares for her young without the help of a mate. The kits grow and develop quickly.

WOLVERINE ENJOYING THE SNOW.
PHOTO COURTESY OF JOHN GOMES AND THE ALASKA ZOO.

WOLVERINE.
PHOTO COURTESY OF JOHN GOMES AND THE ALASKA ZOO.

They are weaned at about 2½ months and begin traveling with their mothers in late spring. The young leave their mother at five to six months.

SOCIAL STRUCTURE: Wolverines live a solitary life, except to mate and care for their young.

DISTRIBUTION: Wolverines are found throughout mainland Alaska and also on some islands of Southeast Alaska.

MOVEMENTS AND MIGRATION: Wolverines have large home ranges and move around considerably to locate food and to find mates. They may travel 40 miles (64 km) per day during the day and night. Male wolverines have a home range of about 240 square miles (622 km^2) while females use about 50 square miles (130 km^2). Home-range size varies, depending upon prey availability, prey type, time of the year, and topography. Wolverines may attempt to exclude other wolverines from their home range.

HABITAT: Wolverines are successful in a variety of habitats, but they appear to do best when they have access to forests, open tundra or alpine areas, and areas without people.

PREY: Wolverines are opportunistic and will kill and eat a wide variety of prey, even prey larger than themselves, such as caribou. However, they usually only scavenge larger animals such as wolf-killed moose, elk, caribou, Dall sheep and mountain goat. Wolverines will not tolerate other wolverines when feeding on large prey. They may tolerate other wolverines when abundant food sources are available, such as during a eulachon (small oily fish) or salmon spawning event. They also will prey on ptarmigan, hares, voles, squirrels and porcupines. Wolverines are well known for their ability to smell and locate small pieces of carrion, even when it is buried under deep snow cover.

PREDATORY CHARACTERISTICS: Wolverines depend upon larger predators, especially wolves, for much of their food. Wolverines will secure carrion taken by these larger predators and even eat frozen flesh and bones. Their strong jaw muscles help them in this effort, and their sharp teeth are good for slicing and tearing at the frozen carcasses. They will watch for other scavengers, such as ravens and eagles, to help them locate carrion that the birds have found. Wolverine feet are large, acting like snowshoes, allowing them to travel more easily over the snow. Ptarmigan may burrow into the snow for protection at night and during cold spells. Wolverines are able to locate these hidden birds, crash through the snow, and capture one or more birds before they can escape. Wolverines have large canine teeth for securing and killing prey. Their aggressive disposition has given them the reputation of being fearless, and they will confront bears and wolves in an effort to steal prey.

Terrestrial Predators 53

Wolverine have formidable teeth and strong jaw muscles.
Photo courtesy of John Gomes and the Alaska Zoo.

Wolverine also have formidable claws.
Photo courtesy of John Gomes and the Alaska Zoo.

WOLVERINE.
Photo courtesy of John Gomes and the Alaska Zoo.

PREDATORS: Trapping by people and predation by wolves and bears act to keep wolverine numbers low. However, most wolverine mortalities are due to starvation.

CURRENT STATUS: Wolverine numbers are naturally low, but their populations are healthy in Alaska. Population densities vary depending upon prey abundance, disease and human presence.

CONSERVATION: Wolverine densities are low in Alaska, and they must cover a lot of ground to find a mate. This problem is somewhat alleviated in that wolverines can mate over a four-month period, May to August. Soon after fertilization, the embryos enter a state of no-growth until early winter when the embryos implant in the uterus and begin to grow. This is called delayed implantation. There are two obvious advantages to using delayed implantation. First, more than one male may breed with the female, so she may deliver offspring with different fathers; this may increase the diversity and survivability of the offspring. Second, delayed implantation regulates the number of young the female has, based on her physical condition. If she has had a good summer, found lots to eat, and was not injured, she may produce as many as four young. If her condition is poor she will likely produce one or no young. This reduces the stress on the female and improves her chances of surviving so that she can reproduce the following year.

Wolverines are highly valued for their beautiful fur. The hair of a wolverine is resistant to frost formation. This makes wolverine fur a practical choice for the ruff of a parka's hood, especially in cold, windy coastal areas.

The wolverine is a symbol of Alaska's wilderness. Protection of wolverines requires a focused effort of understanding and managing both the human harvest and the areas essential to their survival. Wolverine population dynamics appear to be closely correlated to the health of wolf populations. This is a result of the scavenged prey that wolverines secure from wolf kills. Since wolverines use large areas, managing them is difficult. For instance, since wolverines from many adjacent home ranges travel on the same paths, if traps were to be set in just one area of that path, it is likely that ALL the wolverines for hundreds of square miles would be wiped out. Protection likely will require creating refugia of critical feeding and travel areas and locations where wolverines den and raise their young. These areas need to be relatively free from human activities, including over-trapping and wolf control. Wolverines are an indicator of a pristine and wild land.

WOLVERINE RELAXING IN THE SNOW.
PHOTOS COURTESY OF JOHN GOMES AND THE ALASKA ZOO.

AVIAN PREDATORS

BALD EAGLE

The bald eagle (*Haliaeetus leucocephalus*) is the national emblem of the United States of America and is found on the national seal and monies. Bald eagles are in the Falconiformes an order that includes falcons and hawks.

SIZE: When viewed up close, bald eagles appear to be as large as a big dog, especially when they spread their nearly seven-foot (2.1 m) wings. However, large female bald eagles weigh only about 15 pounds (6.8 kg) and the males weigh approximately 13 pounds (5.9 kg).

COLOR: Juvenile eagles start with dark feathers (sometimes with light tan markings), dark beaks and dark eyes. They acquire the characteristic white head and tail by the time they reach five years of age. An adult eagle has yellow feet and beak, and yellow to silver eyes.

SPEED: In a dive, bald eagles may obtain speeds of 100 miles per hour (160 km/hr). They flare their wings and slow before striking their prey or snatching a fish from the water.

LONGEVITY: Bald eagles live 20–30 years in the wild.

REPRODUCTION: Bald eagle nesting trees are usually the larger, older trees in an area, often within sight of water. In Alaska, cottonwood, spruce, cedar and hemlock are the preferred tree species used for nesting. When trees are not available, bald eagles will nest on the ground, on cliffs or on small islands. Ground nesting is mandatory in some regions of Alaska such as the Aleutian Islands, which are treeless.

Nests are built or refreshed with green branches, usually in February to March. Bald eagle nests can be extraordinary structures of sticks, weighing as much as a ton. The nest often is lined with softer materials such as mosses. Nesting trees often succumb to this added weight and crash to the ground. Perhaps this is one reason why bald eagles sometimes have several nests in their territory.

Mating occurs in March to April. The male mounts the female for a quick interlude, lasting only a few seconds, though this may occur many times. Bald eagles lay

BALD EAGLE.
PHOTO COURTESY OF THE US FISH AND WILDLIFE SERVICE.

BALD EAGLE.
PHOTO COURTESY OF JOHN GOMES AND THE ALASKA ZOO.

58   Alaska's Predators

A FEW BALD EAGLES HAVE BEEN SEEN WITH PATCHES OF WHITE OTHER THAN ON THEIR HEADS AND TAILS. THIS ALL-WHITE BALD EAGLE, SEEN IN SOUTHEAST ALASKA, WAS NOT AN ALBINO AND APPEARED TO BE DOMINANT OVER OTHER BALD EAGLES.
PHOTO BY BRUCE WRIGHT.

BALD EAGLE LANDING ON ITS NEST.
PHOTO COURTESY OF THE US FISH AND WILDLIFE SERVICE.

one to three eggs, dull white in color, oblong, and about three inches (7.6 cm) long. Incubation is 35 days in duration and begins as soon as the first egg is laid. The eggs hatch in the order they were laid, so the youngest chick normally is smaller than the oldest chick. This may spell disaster for the younger chick(s) if prey becomes scarce. If the adults cannot bring enough food to the nest, the older and larger sibling(s) will compete for and get most or all of the food. The younger chick(s) can starve or even be pushed from the nest.

After hatching, the young are brooded by the adults to protect them from exposure to wind, rain and sun. As the young get older, the adults spend less time at the nest and more time hunting. The young birds leave (fledge) the nest about three months after hatching. After the young fledge, they are on their own. In the fall the young eagles often find and eat spawned and dead salmon or they will feed on the carcasses left behind by feeding bears.

Bald eagles usually remain paired for life. If a newly established pair fails to raise young after a year or two of nesting attempts, one of the birds may find another mate. If a mate dies, the survivor selects a replacement.

SOCIAL STRUCTURE: Prime nesting sites must be defended or new birds will move in. However, bald eagles may feed together on abundant sources of food. Late fall and winter concentrations of spawning salmon attract eagles from great distances. During the winter up to 4,000 bald eagles can be seen in the Chilkat River Valley feeding on a late chum salmon run.

Spring concentrations of eagles can be found at several locations in Alaska: Copper River Delta, Berners Bay, Sitka Sound, Kenai Peninsula, and the Stikine River, where energy-rich forage fish (herring and eulachon) spawn.

DISTRIBUTION: Bald eagles occur throughout much of Alaska, except for the most northern areas and higher elevations where summers are too short to raise young before the lakes and rivers freeze.

MOVEMENTS AND MIGRATION: If food is available, adult bald eagles tend to stay near their nesting areas all year, protecting the sites from other eagles. During the winter, the interior lakes and rivers freeze, forcing most interior bald eagles to migrate to the coast where they feed on fish. Some bald eagles migrate as far south as Washington and Oregon during the winter, but these are usually the younger birds. Juvenile bald eagles have larger wings than adults. Although juvenile eagles' skeletons are not larger, their feathers are as much as 20 percent larger. This adaptation aids the juveniles' long migrations.

HABITAT: Alaskan bald eagles are almost always found near open water. In Alaska, ocean coastlines are their preferred habitat.

PREY: Prey availability has much to do with what eagles eat. When spawning salmon are available bald eagles will feed almost exclusively on this easily obtainable food. During years when snowshoe hare numbers are at their peak, interior bald eagles may catch and consume these prey items. Bald eagles are known to take larger prey, also. They have been seen taking adult great blue herons, geese and swans. Eagles sometimes take deer that have been forced down to the beach due to deep winter snow. Young seals and otters sometimes fall prey to bald eagles, and in the towns eagles may take domestic prey such as cats and small dogs.

PREDATORY CHARACTERISTICS: Bald eagles have very acute eyesight touted to be five times better than that of humans. This superb eyesight is important for locating prey. To capture their prey, eagles use their power-

BALD EAGLE. *PHOTO BY BRUCE WRIGHT.*

BALD EAGLE. *PHOTO COURTESY OF JOHN GOMES.*

BALD EAGLE.
PHOTO COURTESY OF JOHN GOMES.

ful feet and talons (claws). The talons, three in front and one facing back, pierce vitals and kill their prey. The gripping power is so great that even bone can be pierced and broken. The bottoms of the feet have rough scales for holding slippery fish. The impressive hooked bill is used to tear flesh from the prey.

PREDATORS: Ravens may take bald eagle eggs and chicks. Bears, wolves and foxes sometimes prey upon ground-nesting bald eagles. Intraspecific aggression does occur, and may result in one or both eagles' deaths.

Prior to statehood in 1959, Alaskan bald eagles were thought to be important predators of fox (at fox farms) and salmon. This encouraged the Alaska legislature to begin a predator-control program by establishing a bounty on bald eagles. During the bounty years, 1917–1953, over 128,000 bald eagles were killed and their feet were submitted for the bounty. Alaska's bald eagle numbers have increased dramatically since the termination of the bounty. Today, human technology (power lines, traps, aircraft, contaminants and guns) still causes many bald eagle deaths.

CURRENT STATUS: Bald eagle populations in Alaska are at pre-bounty numbers, approximately 50,000 birds statewide. Some areas are so saturated with bald eagles that some adults cannot find nest sites. Southeast Alaska has the greatest densities of bald eagles; on some of Southeast Alaska's islands, nests are found about every mile (1.6 km) along the shorelines.

ECOLOGY/CONSERVATION: Large, old growth trees are the preferred bald eagle nesting location. These older trees are large enough to carry the weight of an eagle nest, and they usually have openings in their branches allowing easy access for the eagles. Old growth timber (ancient forests) is also the most valuable for the timber

THIS BALD EAGLE WAS FOUND DEAD IN ALASKA'S COASTAL RAINFOREST AND REVISITED MONTHLY TO DOCUMENT ITS DECOMPOSITION. IT HAD DECOMPOSED FOR TWO YEARS WHEN THIS PHOTO WAS TAKEN. PHOTO BY BRUCE WRIGHT.

BALD EAGLE.
PHOTO COURTESY OF JOHN GOMES AND THE ALASKA ZOO.

industry. Bald eagles are protected by the Bald and Golden Eagle Protection Act, which authorizes the U.S. Fish and Wildlife Service to protect eagle nesting sites. Timber operations in Alaska are supposed to leave a buffer of trees around eagle nest trees, or they risk fines and possible imprisonment. Protection of old growth forests near lakes, streams, rivers and ocean shorelines is critical for bald eagle nesting success.

Contaminants such as DDT, lead and oil have caused many bald eagle deaths. After World War II, DDT was used to control mosquitoes. This contaminant began to permeate and accumulate in the food web. DDT is lipophyllic (attracted to fats) and becomes concentrated in higher quantities in larger organisms up the food chain as they eat smaller organisms. A DDT dose of one part per billion (PPB) is used to kill mosquitoes. This very small amount could accumulate and concentrate into thousands of PPB in fish, and even higher levels in eagles. This is referred to as "biomagnification." Concentrations can become so high in some birds (those that feed at the top of the food web including pelicans, osprey and bald eagles) that their reproductive systems can be compromised. The near elimination of use of DDT in North America has resulted in a dramatic recovery of bald eagles and other fish-eating birds.

High contaminant levels may also affect people, and DDT was poisoning us as well. The bald eagle was an indicator species signifying that the environment was not healthy for eagles or for people. New research has revealed that fish-eating birds, such as bald eagles, may be redistributing DDT in the environment through their feces, another indicator showing the need to control chemical outputs into the environment.

Bald eagles also may fall victim to lead poisoning when they ingest waterfowl containing lead shot. The use of steel shot to hunt waterfowl may have reduced this problem, but some eagles die from lead poisoning each year when they eat hunter-killed big game. Hunters usually discard the part of the animal containing the wound and parts of the spent lead bullet. Scavengers, including eagles, can consume this discarded flesh and the lead bullet, which can result in death by poisoning. There is no evidence that this is a serious threat to the Alaskan bald eagle population and there are no current plans to eliminate this hazard to eagles and other predators in Alaska.

Bald eagles can be injured or killed by oil spills when the birds attempt to capture and eat oiled prey. An estimated 250 bald eagles died as a result of the *Exxon Valdez* oil spill — approximately 5 percent of the Prince William Sound eagle population.

Eagles can be caught and killed in leg-hold traps set for furbearers (mink, marten, wolves, etc.). Some of these eagles find their way into one of the three Alaska raptor rehabilitation centers where toes or legs may need to be amputated. The Alaska Trappers

Association attempts to educate trappers about techniques to prevent catching bald eagles and other non-target animals.

Bald eagle reintroduction efforts in the lower 48 states have been successful largely because Alaskan birds have been donated to the effort. From 1982 to 1990, 279 Alaskan eaglets were released in California, Indiana, Missouri, New York, North Carolina and Tennessee. The bald eagle populations in these and some adjacent states are recovering as a result of this effort.

Management of healthy bald eagle populations is straightforward. Eagles need protection from human persecution, they need a healthy and contaminant-free food supply, and they need a healthy habitat with safe nesting situations. If we monitor bald eagle populations we will learn much about what is happening with the environment, and whether it is safe for eagles and for people.

Gyrfalcon

The gyrfalcon (pronounced jur'folken) (*Falco rusticolus*) is the largest falcon in the world. Gyrfalcons are in the Falconiformes, an order that includes falcons, hawks and eagles.

SIZE: The female gyrfalcon stands about 24 inches (61 cm) tall and the male about 22 inches (56 cm) tall. The female has a four- to five-foot (1.2–1.5-m) wingspan and weighs 3–4½ pounds (1.4–2 kg). The male weighs two to three pounds (0.9–1.4 kg). Gyrfalcons have the characteristic long pointed wings of all falcons, but their wingtips do not extend beyond the end of the tail when perched.

COLOR: There are three color phases of gyrfalcons: light, gray or dark, with the female having heavier gray markings on her body and head. The lighter-colored gyrfalcons are more common farther north and in Greenland, with the darker forms more common in the sub-Arctic.

SPEED: Gyrfalcons may obtain speeds of 200 miles per hour (322 km/hr) in a dive. However, they rarely dive on their prey, preferring to use level powered flight. These speeds often exceed 50 miles per hour (80 km/hr).

LONGEVITY: Gyrfalcons can live 25 years in captivity and 12–15 years in the wild.

REPRODUCTION: Males begin defending nesting territories as early as January, and the females arrive in early March. The nest is often a rocky shelf on a cliff that was used by ravens or golden eagles in previous years. Successful nest sites are used repeatedly and can be identified by the streaks of white guano. After pair-bonding displays and copulation, two to seven eggs (usually four) are laid. The eggs are incubated as soon as the first one is laid, to prevent freezing. Incubation lasts 35 days and hatch within one to two days of each other. The chicks are covered in a heavy down. After only 10 days the chicks are able to thermoregulate (maintain their body temperature), allowing the female to join the male in hunting forays. At seven to eight weeks the chicks are able to fly, but they depend upon their parents for food for another month, until mid-September.

SOCIAL STRUCTURE: Gyrfalcons are solitary except during the breeding season when the pair and their young are together until mid-September. They mate for life.

GYRFALCON.
PHOTO COURTESY OF MARK ROBB.

Avian Predators 63

GYRFALCON CHICK IN A CLIFF NEST.
PHOTO COURTESY OF LEE AND SHARI MERRICK.

GYRFALCON.
PHOTO COURTESY OF TRAVIS BOOMS.

TWO YOUNG GYRFALCON.
PHOTO COURTESY OF TRAVIS BOOMS.

(LEFT) LIGHT COLOR PHASE GYRFALCON.
PHOTO COURTESY OF TRAVIS BOOMS.

(RIGHT) GRAY COLOR PHASE GYRFALCON.
PHOTO COURTESY OF TRAVIS BOOMS.

FOUR YOUNG GYRFALCON IN A COTTONWOOD TREE NEST.
PHOTO COURTESY OF TRAVIS BOOMS.

GYRFALCON WITH PREY.
PHOTO COURTESY OF LEE AND SHARI MERRICK.

GYRFALCON WITH PREY; THIS BIRD IS USED FOR FALCONRY.
PHOTO COURTESY OF LEE AND SHARI MERRICK.

DISTRIBUTION: Gyrfalcons are holarctic in distribution and rarely are seen south of 60 degrees latitude.

MOVEMENTS AND MIGRATION: Gyrfalcons are considered non-migratory. However, some birds will disperse south during the winter, sometimes reaching the northern United States.

HABITAT: Gyrfalcons nest in remote regions of the Arctic, and they prefer to use the open tundra.

PREY: Gyrfalcons feed primarily on ptarmigan, but have a large range of prey including geese, ducks, seabirds, ground squirrels, arctic hare and lemmings. Pigeons may be taken in towns during the winter.

PREDATORY CHARACTERISTICS: Gyrfalcons usually wait at a lookout for their prey to appear. Gyrfalcons' keen eyesight gives them an advantage over their prey. Their response to prey is swift. They use the natural features of the terrain for cover and surprise. Unlike the peregrine falcon that uses gravity and speed to strike and kill its prey, gyrfalcons use raw power to chase down the prey and subdue it with powerful feet, or they crush its neck with their large powerful beak. They take prey in the air and on the ground. When

GYRFALCON. PHOTO COURTESY OF MARK ROBB.

large prey is taken, such as an arctic hare, it can be cached and fed on later.

During the winter some gyrfalcons move far enough south to come in contact with human civilization. When in towns and cities they may feed on pigeons; in farming districts they prefer to hunt the open fields and farmland and along the waterfowl flyways.

GYRFALCON.
PHOTO COURTESY OF MARK ROBB.

GYRFALCON.
PHOTO COURTESY OF MARK ROBB.

PREDATORS: Their large size and high-performance flight make the adult gyrfalcon nearly impervious to natural predators. Gyrfalcons can be aggressive when defending their young. They will attack and drive away golden eagles, great horned owls, foxes, wolves, wolverines and grizzly bears, all of which may take the young. Gyrfalcons are not very aggressive to humans, even to researchers who access their nests to collect data. The birds sometimes will fly nearby, calling, but will refrain from striking human intruders.

Some Inuit people use the feathers of gyrfalcons for ceremonial purposes. People take gyrfalcons as chicks (called eyases) to be used in the sport of falconry.

CURRENT STATUS: Gyrfalcon populations, while naturally low, are stable.

ECOLOGY/CONSERVATION: Human activities have had little affect on gyrfalcons — so far. However, the northern ecosystem increasingly is becoming contaminated with chlorinated hydrocarbons such as DDT. These contaminates have found their way into the Arctic marine ecosystem and are concentrated at the top of the marine food web. Some fish-eating marine birds included in the diet of some gyrfalcons may pass their contaminants on to the falcons (see bald eagle ecology section). Migrant prey species may bring contamination up from the south.

An effective way of tracking the health of Arctic habitat is to monitor the top predators. Gyrfalcons are one of the species that should be considered for such research and monitoring programs. By monitoring gyrfalcons we may be able to determine if the Arctic ecosystem deviates from a pristine and healthy condition.

Arctic ecotourism is growing in popularity, and seeing a gyrfalcon can be a high priority for tourists. Good areas to view these large falcons include over-wintering areas with adequate food resources. For those who want a grander adventure, they can arrange to travel to the gyrfalcons' breeding grounds in the Arctic. It is an extraordinary experience to watch this large falcon hunting on the Arctic plain. Ecotourism can have negative effects, so we must monitor this industry as it grows in the Arctic.

GREAT HORNED OWL

The great horned owl (*Bubo virginianus*), also known as the "horned owl," is in the order Strigiformes, containing the other owls.

SIZE: The great horned owl is among the largest owls in North America with ear feather tufts referred to in its name. Horned owls can stand over two feet (0.6 m) tall and have four- to five-foot (1.2–1.5 m) wingspans. Female raptors (hawks, eagles, falcons and owls) are larger than the males. The female great horned owl is about 10–20 percent larger than the male. This sexual size dimorphism may allow the female to hunt different prey in the same territory, thus reducing competition between the pair and increasing the prey species available to them. Other birds of prey also use this strategy. These impressively large and powerful owls weigh two to four pounds (1–2 kg).

COLOR: Great horned owl body feathers are blotched with a variety of browns and tans with some reddish, light-gray and dark bars. The underside is usually lighter gray with dark bars. The feet are feathered to the toes, and the large, black talons are visible and impressive. The great horned owl has large yellow-orange eyes and a facial disc bordered by dark feathers.

SPEED: Horned owls rarely exceed speeds of 50 miles per hour (80 km/hr).

LONGEVITY: Great horned owls have been recorded to live at least 28 years in the wild, while captive birds have been known to live nearly 40 years.

REPRODUCTION: In Alaska, great horned owls will nest in a variety of locations, including tree cavities, cliff ledges and old buildings, but they most often use an old nest of another large bird, such as a raven, hawk or even a magpie. Great horned owls have an advantage for locating the best nests; they begin their nesting cycle in January and February, which is much earlier than their competitors. Once they have established themselves at a nest, even if it belonged to another large predatory bird, the owls are difficult to dislodge from the site. Courtship includes hooting, bowing to each other with drooped wings, and mutual bill rubbing and preening. The male mounts the female for the quick copulation. Eggs are laid every day or two apart, but incubation begins when the first egg is laid to protect it from freezing. Incubation of the two to five eggs takes about 30 days. The chicks grow quickly and are hopping around the nest site at six to seven weeks old. They fledge at approximately 10 weeks of age.

After fledging, the young will continue to be fed for several weeks by the adults. The juveniles disperse in the fall. Adults in northern regions usually migrate south during the winter, but many adults remain near their breeding territories. The juveniles disperse widely. The adults may use the same nesting territory for many years.

Great horned owls aggressively defend their nests, even from humans. Some great horned owl researchers have scars from the talons of an angry and defensive owl.

SOCIAL STRUCTURE: Great horned owls are notorious for their aggressive behavior. A breeding pair will exclude other horned owls and raptors from their nesting territory. Great horned owls mate for life.

DISTRIBUTION: Great horned owls range from North to South America, wherever there are trees for nesting.

MOVEMENTS AND MIGRATION: Great horned owls are residents throughout the year over most of their range, with the exception of many of the most northern owls which will migrate south during the winter. This is especially true during years of low snowshoe hare abundance.

HABITAT: Horned owls are very adaptable birds. They are found in a variety of habitats, including many cities and towns.

PREY: The great horned owl is a very adaptable predator, taking a large suite of prey items. Waterfowl and other birds, squirrels, rats, hares, marten and other small mammals are the prey most often caught in Alaska. Domestic pets are also on their menu, especially cats.

PREDATORY CHARACTERISTICS: Most owls have soft feathers; many feathers have a leading edge that is serrated. These characteristics allow for silent, stealthy flight. Great horned owls hunt by gliding silently through the forest or over meadows. They also like to hunt from perches, usually old dead trees. These hunting techniques give them the advantage of surprise.

GREAT HORNED OWL.
PHOTO COURTESY OF MARK ROBB.

GREAT HORNED OWL CHICKS.
PHOTO COURTESY OF THE US FISH AND WILDLIFE SERVICE.

Nearly silent flight aids in the element of surprise: most of their prey never knows what hit them. The owl will dive and hit the prey hard with its open feet. The talons pierce through the flesh and into vitals and crush the life out of their victims. Smaller prey is eaten whole or ripped into two or three chunks and swallowed. Birds are plucked before being eaten. Larger prey is taken back to the nest or a perch where it is ripped apart with the strong beak and eaten. Some of the prey can weigh three times that of the owl. Since these animals are too heavy for the owl to carry, they are fed upon near the kill site until the owl is satiated. Great horned owls also will hunt walking on the ground to catch small prey, or wading into the water for frogs and fish.

PREDATORS: Natural enemies of great horned owls are other great horned owls and northern goshawks. In both cases, this aggression probably is only during nest site disputes.

CURRENT STATUS: Great horned owl populations are healthy in Alaska. Population densities vary, depending upon habitat and prey abundance.

ECOLOGY/CONSERVATION: Understanding the ecology of great horned owls is the job of the scientist, and this often requires study of the owl's feeding habits. When the owl eats a small vole, bird or mouse whole, the non-digestible fur, feathers and bones are formed into pellets and regurgitated.

Researchers collect the pellets and analyze the contents to determine the owl's prey. Great horned owl pellets are really very interesting to pick through. They can be four inches (10 cm) long and 1½ inch (3.8 cm) thick, and they are usually dark gray to black. If you

Great horned owl.
Photo courtesy of the US Fish and Wildlife Service.

Great horned owl.
Photo courtesy of John Gomes and the Alaska Zoo.

Great horned owl.
Photo courtesy of John Gomes and the Alaska Zoo.

are lucky enough to find a raptor's pellet, make sure you pick it apart carefully and in the presence of some children. You soon will see that youngsters find this experience worth repeating. Try to figure out what the owl was eating by identifying the bones. The skull and teeth are especially revealing for identifying prey species. Great horned owls eat primarily nocturnal prey and that is what you would expect to be represented in the pellets.

Great horned owls are very territorial, and they will not hesitate to attack intruders. To avoid injury most birds avoid fighting, especially with members of the same species. Owls will hoot to keep other owls out of their territories. If the conditions are right the hooting can be heard miles away. When great-horned owl hooting can be heard, especially during the nesting season, this is a warning that all animals should heed.

Owls help maintain a healthy ecosystem by controlling small-mammal population explosions that will result in over-grazing of habitats and the spread of disease. However, all birds of prey are susceptible to human activities, including poisoning from insecticides and other contaminants, loss of habitat and persecution. Do your part to make the environment safe for great horned owls and other predators, and you will make the world a better place for humans, too.

Snowy Owl

The snowy owl (*Nyctea scandiaca*) is in the order called Strigiformes, which includes all the owls. The snowy owl is probably the most striking and distinctive of the world's 146 species of owls, and it is the only white owl in North America.

SIZE: The female snowy owl is larger and heavier than the male, with a wingspan of 4½–5 feet (1.4–1.5 m), a height of 20–27 inches (51–69 cm) and a weight of 3½–4½ pounds (1.6–2 kg). Snowy owls have a large, round head with yellow eyes and black beak and talons.

COLOR: Male snowy-owl feathers are nearly all white with little barring, while the females have white feathers with thick, dark bars.

SPEED: Snowy owls generally fly less than 30 miles per hour (48 km/hr) but they can reach speeds exceeding 50 miles per hour (80 km/hr). These faster speeds are used in defense of their nest, eggs and chicks.

LONGEVITY: Snowy owls live about 10 years in the wild but have survived up to 35 years in captivity.

REPRODUCTION: Snowy owls arrive in their breeding grounds in northern Alaska by mid to late May. The nest is located on a prominent, well-drained mound that provides a commanding view of the surrounding tundra area. The female scratches a hollow in the tundra to protect the eggs from spring winds and snowstorms. Some moss and feathers may be added to the simple nest. The male performs display flights and also displays on the ground. The male mounts the female and mating is quick. All during the nesting season the male maintains a nesting territory with loud hooting and will attack intruding owls, wolves and researchers.

The number of eggs laid, ranging from four to sixteen, is dependent upon the abundance of prey. When prey is less abundant the owls lay fewer eggs. The female lays an egg every two days, and she begins incubating as soon as the first egg is laid. During incubation the male provides all the prey for himself and the female. The eggs begin hatching after 33 days of incubation, in the order they were laid, at two-day intervals. The chicks are accordingly different sizes.

If the food supply should become insufficient, the younger chicks in the snowy owl's nest will be unable to compete with their older siblings. The younger birds may starve or be eaten by their nest mates. The chicks that survive turn from downy white to a fluffy gray,

SNOWY OWL.
PHOTO COURTESY OF JOHN GOMES AND THE ALASKA ZOO.

SNOWY OWL.
PHOTO COURTESY OF JOHN GOMES AND THE ALASKA ZOO.

Snowy owl.
Photo courtesy of John Gomes and the Alaska Zoo.

and at 10 days old turn nearly black. The chicks leave the nest when they are three to four weeks old, well before they can fly. The adults continue to feed the chicks. The young begin to fly (fledge) at about seven to eight weeks old, and they are completely independent at about nine weeks.

Social Structure: Outside of the breeding season, snowy owls usually are solitary.

Distribution: Snowy owls breed in Alaska's Arctic tundra, but during the winter they may be found as far south as California, Georgia and Texas.

Movements and Migration: During the winter some snowy owls stay on the breeding grounds in the Arctic, while others migrate or wander as far south as the northern contiguous United States, with exceptional sightings recorded as far south as California, Georgia and Texas.

Habitat: Snowy owls are synonymous with the open Arctic tundra. During the winter they may migrate south where they use open habitats, including prairies, marshes or shorelines, to find prey.

Prey: Snowy owls can capture and eat a variety of prey items including hare, ptarmigan and other birds, but lemmings are their primary prey. During the winter many snowy owls migrate to areas where prey is available and abundant, including meadow mice (voles), white-footed mice, deer mice, rats, pheasants, fish and carrion.

PREDATORY CHARACTERISTICS: Lemmings are snowy owls' primary prey. Lemming populations are cyclic, and an abundant population can crash in less than a year. One population of lemmings may be very high while a neighboring population may be very sparse. This makes for patchy prey, which predators try to locate. To adapt to the patchy prey, snowy owls are nomadic; they can breed in a different location every year. This allows the owls to nest wherever lemmings are most abundant.

PREDATORS: Snowy owls are very alert and aggressive. They will attack intruders in their nesting territory. However, foxes and wolves do take the chicks. Jaegers, a predatory bird, can take the owls' eggs and chicks. In some areas of Alaska, snowy owls are killed legally for human consumption. A popular technique used is to put a leg-hold trap on top of a post in the middle of the flat tundra. To gain a better hunting vantage point, the owl will land on the post and is trapped easily.

CURRENT STATUS: Snowy owl populations are healthy in Alaska.

ECOLOGY/CONSERVATION: Snowy owls swallow small prey whole. Large prey, such as ptarmigan and hares, are torn apart. Any bones, teeth, feathers and fur eaten are non-digestible; they are compacted and then regurgitated as pellets. Since a snowy owl must eat up to a dozen mice each day, large accumulations of pellets can be located around perches. Biologists interested in owl and small mammal ecology can extract the bones from the pellets to determine the number and type of prey taken by the owls. Understanding the population dynamics of the prey helps scientists to understand the predator.

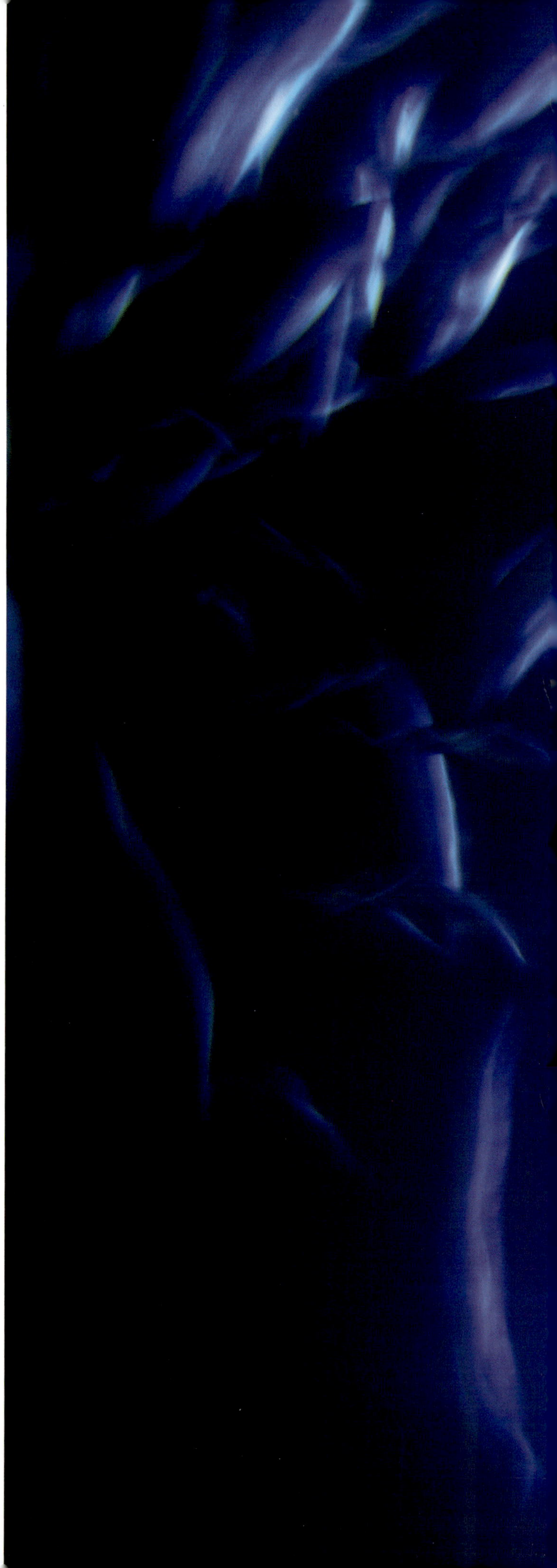

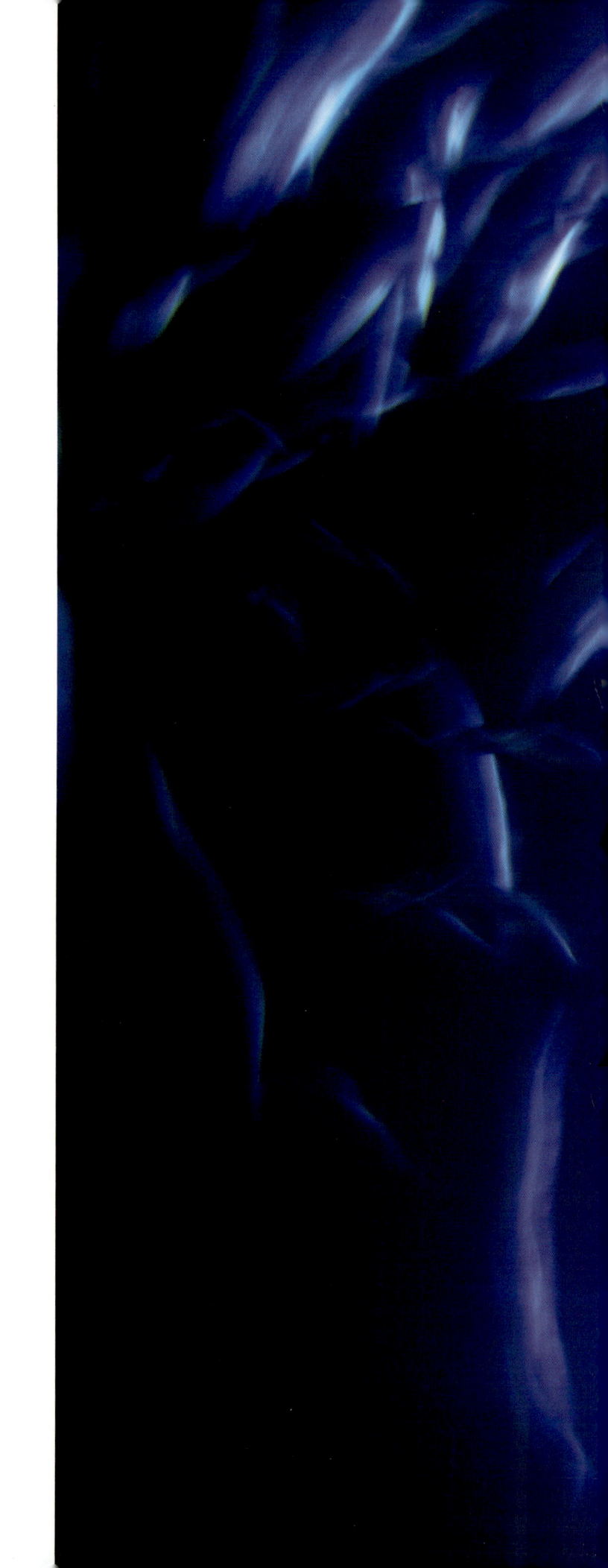

MARINE PREDATORS

SUNFLOWER STAR

The sunflower star (*Pycnopodia helianthoides*) is a very effective, aggressive predator capable of significantly altering marine ecosystems. It is the largest and fastest sea star. Sea stars are in the phylum Echinodermata and the class Asteroidea, which contains 1,500 species.

SIZE: Juvenile sunflower stars initially have only five arms, and the arms are less than 1/8 inch (a few mm) in diameter. Additional arms are added as they grow. Adults can be over three feet (1 m) across, weigh more than 11 pounds (5 kg), and have up to 28 arms.

COLOR: Sunflower stars are orange, red, purple and yellowish in color. Hard, blunt spines (projections of the internal skeleton) cover the upper body surface. The spines may be colorful, too.

SPEED: The sunflower star is the fastest sea star and can move at speeds exceeding 18 feet per minute (10 cm per second). Sea stars usually move slowly, about six inches (15.2 cm) a minute, and often remain in one position on the sea bottom for extended periods of time.

LONGEVITY: Sea stars have no old age and, barring accidents, predation, starvation and disease, they can live a long time. Accordingly, some sea stars may live more than 30 years.

REPRODUCTION: The sunflower star typically spawns between the months of March and July, with the gonads enlarging during winter. Sexes are separate, but they look alike. Sunflower stars will form large aggregations when spawning. Both males and females simply release their gametes into the water. Their eggs and sperm are denser than seawater and form a puddle of eggs and sperm on the sea floor where fertilization of the eggs occurs. A female sea star can produce as many as 2,500,000 eggs each year. While this seems incredibly high, only a few young sea stars survive the very treacherous larval life stage.

SUNFLOWER STAR.
PHOTO COURTESY OF NATIONAL OCEANIC AND ATMOSPHERIC ADMINISTRATION.

THIS SUNFLOWER STAR IS TRAVELING ACROSS THE TOP OF EELGRASS IN A PRINCE WILLIAM SOUND INLET.
PHOTO COURTESY OF MANDY LINDEBERG.

Eggs develop into a microscopic larval stage that is free-swimming. Free-swimming larvae move to, and develop in, the top of the euphotic zone, the ocean's uppermost surface. The microscopic larvae are unable to swim against ocean currents, and they can be transported great distances, aiding their dispersal into new areas. Once development has progressed nine to ten weeks, the small sea stars fall to the ocean bottom and mature to adults. Sea stars also have the ability to reproduce asexually, and they can regenerate lost body parts.

SOCIAL STRUCTURE: *Pycnopodia* can be found in populations made up of many individuals, but they also can be found as solitary individuals. When solitary individuals meet, they may display aggressive behavior towards one another.

DISTRIBUTION: Sunflower stars are found from Unalaska Island, Alaska, to Baja California, Mexico. They can be found in waters that vary from low intertidal shallows to a depth of 1,430 feet (435 m).

MOVEMENTS AND MIGRATION: The majority of sea star dispersal occurs during the larval stage when ocean currents carry them around. Larval sea stars can disperse far from where they were born (hundreds of miles or kilometers). Adult sunflower stars move surprisingly long distances in search of prey (miles or kilometers). They move with the highly coordinated movement of more than 40,000 small arms on their undersides, which are referred to as tube feet.

HABITAT: Sunflower stars use a variety of habitats, from rocky shores to muddy bottoms.

PREY: Sunflower stars eat mussels, oysters, clams, scallops, snails, abalone, crabs, worms, sea urchins, sea cucumbers, shrimp and decaying matter.

PREDATORY CHARACTERISTICS: The sunflower star is an extremely aggressive carnivore. Intertidal animals avoid sunflower star predation in a number of ways; they may have strong shells (clams), sharp spines (sea urchins), use small appendages called pedicellare to snip off the sunflower star's tube feet (sea urchins), teeth (polychaetes) or claws (crabs). Sunflower stars are able to overcome most of these defenses. When touched by the sunflower star, scallops will open and close their valves quickly, allowing them to "swim" away. Abalone, limpets and snails move away at relatively high speeds when they are touched by a sea star. Some prey, such as sea urchins and some snails, also can detect sea stars by scent, initiating the prey to move away.

Although the sunflower star seems slow to our eyes, it is faster than most of its mobile prey species, catching them with its superior speed. Using its thou-

THIS SUNFLOWER STAR (LEFT) AND LEATHER STAR (RIGHT) HAVE BEEN PLACED ON THE PONTOON OF A RESEARCH SKIFF FOR A COMPARISON PHOTOGRAPH. *PHOTO COURTESY OF JOHN HARVEY.*

A RESEARCHER IS PHOTOGRAPHING THIS SUNFLOWER STAR. *PHOTO COURTESY OF THE US FISH AND WILDLIFE SERVICE.*

sands of tube feet, the sunflower star can dig down six inches (15.2 cm) or more to secure clams. While some sea stars digest their food externally, *Pycnopodia* ingests its entire prey. Once the soft tissue is digested, the shells of calms, mussels and other organisms are discarded through its mouth.

The best protective mechanism for potential prey is to live higher in the intertidal zone than sea stars are capable of using. The next time you are on the beach at low tide, look for the zone of predation and the sea stars that maintain that line.

PREDATORS: Many different small predators prey upon larval sea stars. Adult sunflower stars are preyed upon by sea otters, gulls, crows and the sun star *Solaster stimpsoni*. Their primary predators are king crabs and Tanner crabs. Humans often kill sea stars to reduce their predation on commercial and subsistence resources such as oysters and mussels, and sea stars are collected for use as fertilizer. Some sea otters eat only part of a sea star before discarding it. The discarded sea star may survive and regenerate its missing limbs. To escape a predator, a sunflower star may detach an arm or two as a sacrifice for the predator to eat. While the predator is distracted eating the detached arm(s), the rest of the sunflower star may be able to escape.

TUBE FEET. *PHOTO BY BRUCE WRIGHT*.

CURRENT STATUS: Sunflower star populations in Alaska are healthy. Sunflower star populations tend to recover quickly after being reduced by predation or other environmental factors.

ECOLOGY/CONSERVATION: The intertidal zone is located at the boundary between marine and terrestrial ecosystems, between the high and low tides. The tides create conditions that influence the distribution of organisms living in the intertidal zone. The intertidal zone can be subdivided into three additional zones: the upper intertidal zone is submerged only during the highest tides of the year, the middle intertidal zone is submerged during most high tides and exposed during most low tides, and the lower intertidal zone is exposed only during the lowest tides of the year. The intertidal organisms have adapted to the physical environment with its periodic exposure during low tides and with the shock of waves battering against them. The physical environment becomes more hostile higher up the intertidal zone, and the lower limit for many species is defined by predation. The top predator in many intertidal ecosystems is a sea star, and its predatory impact is often responsible for setting the lower distribution limit for several other species.

Marine intertidal ecologists invariably must consider the importance of sea stars in the structuring of the intertidal zone. Dramatic changes in the intertidal zone and in the distribution of sea star prey result when researchers exclude sea stars, using devices such as fences. The eons of time during which these sea stars and their prey have co-existed in the ocean has resulted in interesting relationships. Researchers are beginning to understand some of these complex associations.

For example, the sunflower star, which is a lower-intertidal and sub-tidal species, is an important predator of sea urchins. The predatory sunflower star greatly reduces the abundance of the herbivorous urchin, allowing kelps to proliferate. When the sunflower star populations are low, due to their moving to other areas or due to reductions from predation, sea urchin populations may increase dramatically. Once the sea urchins return, they graze off the marine algae, the kelp beds disappear, and the nearshore ecosystem is altered dramatically. This effect by the sunflower star is referred to as a "trophic cascade."

Giant Pacific Octopus

The giant Pacific octopus (*Enteroctopus dofleini*) is in the mollusk phylum (which includes snails, clams, oysters, mussels and squid) and the family, Octopodidae (the benthic octopuses). Octopuses also are impressive and successful predators. Octopuses may be the most intelligent invertebrates. Among all octopuses, the giant Pacific octopus stands out for its large size.

SIZE: The largest giant Pacific octopus ever recorded had an estimated weight of 600 pounds (272 kg) and had an arm span exceeding 30 feet (9.1 m). These larger records are not well confirmed. In particular, the size and arm-span of the "largest ever" likely were estimated underwater and may be exaggerated. There is a well-substantiated record from British Columbia of an animal of 156 pounds, measuring almost 23 feet from arm-tip to arm-tip. Larger estimates are anecdotal and suspect. However, most giant Pacific octopus females reach about 55 pounds (35 kg) while males get to be about 90 pounds (41 kg).

COLOR: An octopus can change the color and texture of its skin to match its surroundings, allowing some protection from predators. However, the giant Pacific octopus is generally pinkish-orange in color. An octopus can also conceal its escape by squirting a cloud of black ink.

SPEED: The giant Pacific octopus sucks in water over its gills and out a tube called the "siphon." The octopus can force water through the siphon at an increased rate to jet the animal backwards a few miles per hour (5 km/hr), and is another means of escaping predation. Large-scale movements in the wild may depend on riding the tides and do not represent the animal's swimming speed through water.

LONGEVITY: Most octopus species live one to two years, but in captivity the female giant Pacific octopus lives about 3½ years and the male lives up to five years.

GIANT PACIFIC OCTOPUS AS VIEWED FROM A SUBMERSIBLE.
PHOTO COURTESY OF DAVID SCHEEL.

GIANT PACIFIC OCTOPUS HUNTING FOR PREY.
PHOTO COURTESY OF DAVID SCHEEL.

GIANT PACIFIC OCTOPUS IN ITS ROCK DEN.
PHOTO COURTESY OF DAVID SCHEEL.

REPRODUCTION: When ready to mate, the female releases a chemical attractant that guides males to her hiding place. If more than one male is attracted, they will fight for the opportunity to mate with the female. If the female accepts the male, he will transfer a couple of packets of sperm, using the end of his third arm. The third arm is modified to do this and has no suckers on its end. The female stores the sperm for later use. The male leaves to search for other females.

The female may spend up to a month selecting a deep-water den where she lays her eggs, usually under a large rock. She pulls in additional rocks to seal off all the entrances to the den. She lays one egg at a time, fertilizing it as it passes through her reproductive tract, out her siphon, and onto a sucker. The eggs, about the size of a grain of rice, are attached to the den's ceiling in strings of about 200 eggs at a time. During a three-week period she may lay more than 100,000 eggs.

The eggs develop and incubate for 6½ months, during which time the female — who neither leaves the den, nor feeds — keeps them free of bacteria, algae and other organisms that may harm them. She also blows water across the eggs to keep them oxygenated.

After 6½ months of constant attention by the female, the eggs begin to hatch. They hatch at night to avoid predation. The female blows water to help the baby octopuses, called paralarvae, emerge so they can swim to the surface of the ocean. Soon after the young hatch, the female will die.

The paralarvae spend several weeks at or near the surface of the water, the top of euphotic zone, where many small prey items and other zooplankton gather to feed. If the paralarvae survive the weeks in the euphotic zone they become heavy and descend to the bottom of the ocean where they remain for the duration of their lifetimes.

DISTRIBUTION: The giant Pacific octopus is found throughout the North Pacific Ocean, from the shallows of intertidal pools down to hundreds of feet deep.

MOVEMENTS AND MIGRATION: As paralarvae, the small octopuses are at the mercy of the winds and currents, which may transport them many miles from where they hatched. Adult giant Pacific octopuses can travel for miles along the ocean bottom in search of food, mates and suitable dens.

HABITAT: Giant Pacific octopuses use the upper part of the ocean (top of euphotic zone) when they are young, and they use the bottom when they are adults. The adults prefer to have a den that is secure from predators. Larger, deeper animals actually seem to depend less on dens than do smaller, shallower ones. Work in Japan indicates that big, deep animals may actually feed in the open in areas without dens. However, when so exposed, they prefer dens if available, which makes the lair-pot, a trap resembling the animal's lair or cave, fishery in those areas successful.

PREY: The giant Pacific octopus starts life as a small creature feeding on smaller zooplankton at the surface of the water. When octopuses first settle to the bottom they feed on amphipods, a group of small invertebrates, and other small crustaceans and bivalves. As they get larger, the giant Pacific octopuses eat crustaceans such as crab and shrimp, and mollusks, including clams and snails. They also eat fish.

PREDATORY CHARACTERISTICS: The giant Pacific octopus is a well-equipped predator. Its head contains an advanced brain, mouth and two large eyes, and its mantle contains internal organs. The eyes of an adult giant Pacific octopus are about the same size as an adult human's eyes. Giant Pacific octopuses have good vision in the dark, and they feed mostly at night. The octopus's eight arms are equipped with as many as 2,200 suckers arranged in two rows. It will seize its prey and, in the case of crab, wrap its arms around the crab to control and dismember it.

Octopuses open their well-protected prey using three methods. For some prey (e.g., scallops) they simply can pull the shell open with their suckers. For more difficult prey whose shells are still thin, they can chip or break the shell using their beak. The most difficult prey is drilled, using a combination of salivary enzymes that soften the shell and the scraping radula, a special mechanism with tiny teeth, located near the mouth, to wear a tiny hole in the shell. A toxin is injected through the hole that then allows the octopus to pull open the prey. Variations on these themes also include snapping the spinal cord of fish. If at low tide one comes upon an octopus den, one may notice some of these shells with oval pinholes amongst the midden of shells outside the entrance, evidence of the octopus's handiwork.

PREDATORS: Octopuses have many predators throughout their development, and adults must be wary or risk being eaten by sharks, halibut or marine mammals. In addition to a suite of natural predators, people also harvest octopuses for food and for fish bait. Commercial halibut fishermen refer to octopuses as "halibut candy," and they like to use them for bait because they stay attached to the hooks. Most divers cherish the times they encounter a giant Pacific octopus. Some divers, who also like to eat octopus, will cut off a single arm for dinner. The octopus likely survives the loss of an arm, which can regrow, and probably hopes the next diver is more compassionate.

CURRENT STATUS: Not much is known about population numbers of giant Pacific octopuses, so it is difficult to determine if their populations are healthy or not. They appear to be numerous in undisturbed areas and in the right habitats. A commercial fishery in Japan appears to be operating sustainably, while a small fishery in British Columbia did not, and was shut down.

ECOLOGY/CONSERVATION: The top of the euphotic zone contains many very tiny marine plants, called phytoplankton and the small grazers, zooplankton that feed on these small plants. The young of octopuses, crabs, sea stars, clams and others use the top of the euphotic zone during their larval stage. Contaminants, such as petroleum products from boats and jet skis and oil from spills, can cover and poison the microlayer (very top of the euphotic zone). Billions of phytoplankton and zooplankton, including octopus larvae, can perish from human-derived pollutants without anyone ever noticing. Protection of the microlayer is extremely important if these organisms are to thrive in the future. In Alaskan waters, juvenile octopuses take up residence in shallow-water kelp beds and intertidal beaches. As with all other organisms in these habitats, they are at risk of being damaged in the event of an oil spill.

Checks and balances in the ecosystem have been incorporated in the evolution of prey and their predators. However, when an invasive species (introduced non-native species) is released into the system, the indigenous species are at risk of exaggerated population increases or decreases, created by a cascading effect of new predator–prey relationships, and competition with the introduced species. Alaska's harsh environment and a strict policy adopted by the Alaska Department of Fish and Game have reduced the occurrences of introduced

GIANT PACIFIC OCTOPUS EATING A CRAB.
PHOTO COURTESY OF DAVID SCHEEL.

GIANT PACIFIC OCTOPUS AS VIEWED FROM A SUBMERSIBLE.
PHOTO COURTESY OF DAVID SCHEEL.

species in Alaska. However, oil tanker traffic into Prince William Sound has helped establish several invasive species into the marine ecosystem. If these species become abundant, as has happened in many aquatic ecosystems (e.g. Great Lakes, San Francisco Bay) they could exert a bottom-up control of the marine ecosystem that could drive a regime shift, changing the hot spots (species composition or locations) and altering the keystone species and dominant predators.

SALMON SHARK

Salmon sharks (*Lamna ditropis*) are fish, in the *Lamnidae* family of sharks. This family includes the great white and mako sharks. *Lamnidae* sharks are warm-blooded (partially endothermic) and salmon sharks are the warmest of the *Lamnidaes,* as warm as 20°F (7°C) warmer than the waters in which they swim.

SIZE: Salmon sharks can grow to about 12 feet (3.7 m) and weigh over 1,000 pounds (454 kg). Females are slightly larger than males.

COLOR: Salmon sharks are dark gray to nearly black above. The underside is whitish with gray blotches. The shark's colorations help camouflage it both from above and below.

SPEED: Salmon sharks are well adapted to travel swiftly through the water. Their bodies are stocky with a pointed or conical snout. Salmon sharks use their large, powerful tail to propel them through the water. Another adaptation for speed is their rough skin. The skin holds water, creating a water-to-water low-resistance interface. When doing submarine research, the U.S. Navy became interested in how salmon sharks could travel so fast in water. Navy researchers reported clocking salmon sharks at over 50 miles per hour (80 km/hr), which puts them among the fastest of fish.

LONGEVITY: Most fish are aged using bony structures called otoliths; however, sharks do not have bones, making it difficult to age them. A recent discovery has found that the snout of the salmon shark is ossified, or bone. What appear to be annual rings in the snout's bone indicate that a 6½-foot (2 m), 300-pound (136 kg) female salmon shark is about 10 or 11 years old. From this information, researchers estimate that salmon sharks may live 25 years or more.

REPRODUCTION: Male salmon sharks become sexually mature at about five years and females at about eight to 12 years. Breeding takes place during the fall. Salmon sharks are ovoviviparous, meaning that they produce eggs that hatch within the female's body. During gestation the young sharks attack and consume non-developing eggs. This is common among sharks. The female, therefore, gives birth to two to five live young, called pups. The estimated gestation period is around nine months to a year. Young are 32–34 inches (81–86 cm) long at birth and are fully equipped with sharp senses and sharp teeth, the better to take prey and avoid predators.

SOCIAL STRUCTURE: Salmon sharks feed together on schools of salmon and other fish and may work together to herd the prey. The sharks appear to be non-aggressive towards each other as they feed. During breeding, males will bite on to females to hold on. What appears to us to be dangerous and aggressive behavior seems to work for sharks. The sharks seem to not be troubled by pain.

DISTRIBUTION: Salmon sharks can be found in near-shore and offshore waters of the North Pacific Ocean, from Baja California to the Bering Sea and west to Russia, Korea and Japan.

MOVEMENTS AND MIGRATION: Satellite tags attached to salmon shark dorsal fins have revealed that in the summer some sharks migrate to near-shore waters in pursuit of prey. They tend to use deeper and more offshore waters after salmon spawning ends in October. Salmon sharks are highly migratory and may move thousands of miles (>10,000 km) each year in search of prey. Some of the salmon sharks in Prince William Sound, Alaska stay in these waters year round and feed on the large schools of herring. In Russia, salmon sharks are called herring sharks.

HABITAT: Salmon sharks can be found patrolling nearshore waters as they search for prey. However, they spend much of the year in the open ocean, often at great depths. During the summer, salmon sharks hunt near the thermocline, an area in the water column where there is an abrupt change in water temperatures.

PREY: Salmon sharks are opportunistic feeders. They consume a wide variety of prey including salmon, squid, rockfish, pollock, herring, capelin, sablefish, mackerel, sculpin, tomcod, daggerteeth, lantern fishes, pomfret, shrimp, lancet fish, spiny dogfish sharks, arrowtooth flounder and sea otters.

PREDATORY CHARACTERISTICS: Salmon sharks are equipped with good vision and sense of smell to aid in locating and attacking prey. Another well-developed sense is their ability to detect weak electromagnetic fields that are emitted by the muscles of swimming fish and other prey. Sharks have a number of large pores, or channels, in their snouts that are used to detect these weak electrical fields. Sharks' "sixth sense" is so acute as to allow them to track their prey by following the wake the prey leaves in the water.

Muscles work more efficiently when warm. Most fish lose their internal heat to the surrounding water. A salmon shark's warm-bloodedness is achieved by countercurrent heat exchange in which heat produced by internal muscle activity is used to warm the oxygenated blood returning from the gills. Salmon sharks benefit from warm-bloodedness and more efficient muscles by being able to reach higher swimming speeds. Salmon sharks also are equipped with several rows of moderately large, smooth-edged teeth. These are used to grasp and tear their prey into bite-sized pieces.

Much of the salmon shark's prey is taken in deep water. However, some salmon sharks will move into shallow bays and the mouths of salmon streams to pursue salmon that are preparing to spawn. The sharks concentrate their efforts in these hot spots, and as many as 1,000 salmon sharks per square mile (386 salmon sharks per km^2) have been observed in these areas. The sharks will hunt the salmon in groups, or packs, similar to how wolves hunt their prey. Salmon

THIS SALMON SHARK WAS MEASURED AND TAGGED AND IS READY FOR RELEASE. *PHOTO BY BRUCE WRIGHT.*

SALMON SHARKS HAVE DARK BLOTCHES ON THEIR UNDERSIDE. *PHOTOS BY BRUCE WRIGHT.*

are attacked from below or behind. Salmon sharks have been seen leaping out of the water in pursuit of their prey and may clear the water with a salmon in their jaws.

At least two accounts of salmon sharks taking sea otters were reported in Prince William Sound, Alaska. In one account the 300–400-pound (136–181 kg) shark attacked a female otter as her pup swam nearby. The shark grabbed the sea otter and shook, tearing the otter to pieces only 10 feet (3 m) from a fishing boat. Otter blood was splattered across the boat's stern and onto the fishermen. The pup sea otter escaped, but likely died of starvation or was preyed upon by bald eagles, which are on the lookout for unattended young otters.

Salmon sharks are considered dangerous because of their large size and aggressive nature. However, they are rarely aggressive towards people, and there is only one account of a salmon shark biting a person.

PREDATORS: Other sharks may prey upon salmon sharks and they are sometimes taken by orcas. It is not known if the transient type and/or resident type orcas are taking the sharks. Some commercial salmon fishermen regularly kill salmon sharks, and there is an active and growing sport fishery in Alaska for salmon sharks.

CURRENT STATUS: Japanese scientists estimate more than 2,000,000 salmon sharks hunt the North Pacific Ocean, but this estimate is controversial and may be high. Their population appears to be healthy in Alaska. However, salmon sharks are a long-lived, slow-grow-

BRUCE IS POSITIONING THIS SALMON SHARK SO IT CAN BE MEASURED, TAGGED, AND RELEASED. PHOTO BY BRUCE WRIGHT.

AUTHOR HOLDING SALMON SHARK JAWS.
PHOTO BY TOM CAMPBELL.

THIS MEASURED AND TAGGED SALMON SHARK IS READY FOR RELEASE. *Photo by Bruce Wright.*

ing species with a low reproductive rate. This makes them susceptible to over-exploitation. Fisheries on sharks would likely result in a dramatic decline in the regional shark population, similar to what has occurred to other shark populations in other oceans. Alaska fisheries managers are taking a conservative management approach to shark fisheries. This probably has helped in maintaining healthy shark populations in Alaska and also a relatively stable marine ecosystem. The Alaska fisheries managers have not addressed as yet the many fishermen in the region who regularly kill any sharks they catch by cutting off their tails or shooting them on sight.

ECOLOGY/CONSERVATION: Salmon sharks move into areas of high food abundance, or "hot spots." They consume the schools of fish until the amount of prey is reduced or dispersed. Then the sharks move on to the next patch of food. Salmon sharks have learned where and when these hot spots occur. This behavior may result in salmon sharks intercepting salmon runs, but they take far less than the number of fish harvested by fishermen.

An interesting thing about salmon shark predation in Alaskan waters is that it may afford some stability to the ecosystem. Mathematical models show that if a predator-control program removed these sharks from the ecosystem some marine species populations would decline. Scientists think it would work like this: If the sharks were removed, several fish species numbers would increase, especially arrowtooth flounder. The increased flounder population would remove smaller forage fish such as capelin and herring, which reduces the food supply for other fish, seabirds and marine mammals. Some marine mammal populations in the Gulf of Alaska are already at critically low numbers, so further declines would have a devastating effect on the ecosystem and likely would force more fishing restrictions to protect the food of the marine mammals. The complexities of the predator–prey relationship and their effects on the marine ecosystem are difficult and next-to-impossible to predict and manage. A precautionary approach to fisheries and predator management is the wisest choice.

PACIFIC SLEEPER SHARK

The Pacific sleeper shark (*Somniosus pacificus*) is a secretive, deep-dwelling shark referred to by many fishermen as the "mud shark." The Greenland shark is found in the North Atlantic Ocean and is likely the same species but with different common names. Perhaps both of these sharks should be renamed as the polar shark.

SIZE: Pacific sleeper sharks have been caught that exceed 20 feet (6 m) in length, which would weigh in at about 8,000 pounds (3,600 kg). This is approaching the size of adult orcas.

COLOR: Pacific sleeper sharks are uniformly dark gray to black with round fins and tail. Their eyes are small and black.

SPEED: Sleeper sharks are generally sluggish and normally swim slowly. They probably rarely exceed speeds of a few miles per hour (5 km/hr).

LONGEVITY: Pacific sleeper sharks probably live more than 40 years. This age estimate is based upon the size this species obtains and upon the average growth rates. Determining the age of sharks is problematic. Bony fish can be aged by counting the annual rings on a bone in the ear called an otolith; most sharks do not have any bones and no shark has an otolith.

REPRODUCTION: There is little information on Pacific sleeper shark reproduction. Only recently have scientists learned that sleeper sharks bear live young.

SOCIAL STRUCTURE: Deep underwater video has captured many Pacific sleeper sharks feeding together on whale carcasses. The sharks appear to be non-aggressive towards each other as they feed. No information is available on other sleeper shark social interactions.

DISTRIBUTION: Pacific sleeper sharks are found in polar and sub-polar waters throughout the year. They occur in the Pacific Ocean from Baja California north to the Bering Sea, Chukchi Sea, Beaufort Sea, and to the Okhotsk Sea off of Japan. They inhabit cold, deep waters to depths exceeding 6,500 feet (1,981 m). At higher latitudes sleeper sharks use shallow as well as deep waters.

MOVEMENTS AND MIGRATION: Satellite tags have been attached to Pacific sleeper sharks to determine their movements. Tag data indicates that individual sharks moved from the bottom, 2,000 feet (610 m) deep, to the surface each night, apparently to feed. The limited tag data indicate that they did not migrate to other areas.

HABITAT: Pacific sleeper sharks use the ocean bottom for resting and to feed on fish and large, sunken prey such as dead whales. At night they come to the surface to feed.

PREY: Pacific sleeper sharks have a huge gut capacity and can fill it with copious amounts of food. Of 33 sleeper sharks sampled on an International Pacific Halibut Commission survey, five of the sharks had empty stomachs and 28 contained prey including salmon, squid, cod, pollock, squid and octopus beaks, harbor seal and other marine mammal tissue. One shark had nine adult pre-spawning chum salmon in its stomach. Sleeper sharks have been documented feeding on dead whales, gorging themselves on blubber and flesh.

PREDATORY CHARACTERISTICS: Pacific sleeper sharks are designed for stealth. Their eyesight is probably poor, but good eyesight is not necessary since they have an exceptional "sixth sense" to detect very slight electromagnetic fields. Muscle activity, even the beating of an animal's heart or the movement of its diaphragm, emits an electrical signal that the sharks use to detect, locate and attack their victims. The electromagnetic signals guide the shark right in.

For sleeper sharks, darkness is not a deterrent to detecting prey, but instead a cover or camouflage. Under cover of darkness, prey would be less likely to detect a sleeper shark coming up from the depths.

Pacific sleeper shark teeth are quite different in the lower jaw compared to the upper jaw. The upper jaw has small, sharp, conical teeth much like those in halibut. These are used to seize and hold prey. The teeth in the lower jaw are interlocking, forming a serrated blade used for slicing. Sleeper shark bite marks resemble large three-quarter moons or slices.

Sleeper sharks attack suddenly and without warning. A harbor seal might be floating on the surface of the ocean trying to catch its breath when it is attacked from below by a 400-pound (181 kg) sleeper shark. A bite to its midsection and the seal is eviscerated and struggling for its life. At minimum, the shark gets a large chunk of skin and blubber, likely enough to cause the seal soon to die. However the shark will finish the job by ripping the seal to bits, eating and digesting the entire animal. Smaller animals such as adult chum salmon or black cod usually are eaten whole.

Halibut and black cod struggling on a fisherman's long line (bottom-set line with hundreds of baited hooks) attract sleeper sharks. The struggling fish emit signals that the sharks can detect from long distances. The sleeper sharks bite chunks out of the halibut. When sharks try to eat the cod whole, the same hook that

84   Alaska's Predators

Pacific sleeper shark caught on a longline and landed on the beach for sampling.
Photo Mark Wright.

This Pacific sleeper shark was caught in a bottom trawl in the western Bering Sea.
Photo courtesy of Alexei Orlov.

This Pacific sleeper shark is being dissected by Dr. Orlov to determine its diet and reproductive status.
Photo courtesy of Alexei Orlov.

Pacific sleeper shark.
Photo courtesy of Elliott Hazen.

This Pacific sleeper shark was caught in a bottom trawl in the eastern Pacific Ocean.
Photo courtesy of Dave Wagenheim.

caught the cod may hook the sharks. The struggling sleeper sharks tangle and damage commercial fishing gear, forcing many fishermen to change fishing areas.

PREDATORS: Pacific sleeper shark tissue is reported to be toxic to people and other animals. They are probably only preyed upon by other sharks. Commercial fishermen regularly kill Pacific sleeper sharks that they catch while fishing for black cod and halibut.

CURRENT STATUS: Pacific sleeper shark numbers increased dramatically in the North Pacific during the 1980s and 1990s. In areas where few sharks ever were caught before, now fishermen are catching many more sleeper sharks. Many fishermen are reporting more and larger sharks each year in the Gulf of Alaska.

ECOLOGY/CONSERVATION: During the 1970s, temperatures in the North Pacific Ocean increased, followed by a change in species composition in the region. The ecosystem supported great quantities of shrimp and crab before the 1970s, but these species nearly disappeared and were replaced by pollock, cod, halibut and arrowtooth flounder. This species composition change is referred to as a "regime shift." Other noteworthy and dramatic changes included decreases in sea lions, seals and forage fish (capelin, sandlance and herring) and increases in salmon sharks and Pacific sleeper sharks.

One theory explaining the regime shift involves increased winds, global warming, the Gobi Desert, iron and a group of phytoplankton (small single-celled plants) called "diatoms." As the earth warms due to global warming, scientists predicted, and have seen, stronger and more persistent winds. When these winds are especially strong they can sweep across the Gobi Desert of Mongolia carrying tons of dust laden with the element iron. Much of this iron is deposited in the North Pacific Ocean, which promotes the growth of a class of organisms called diatoms. Diatoms are single-celled plants and can grow rapidly if the conditions are right. The diatoms use iron in a process that keeps them near the water's surface and in the euphotic zone where they capture the sun's energy. When the iron supply is used up, the diatoms sink to the bottom of the ocean. If there is lots of iron and it comes in a steady supply, the diatom population blooms and diatoms stay near the surface of the ocean and promote the surface food web and ecosystem.

But if the winds are sporadic the diatoms grow, but soon deplete the iron in the water and sink to the bottom. This may be better for the ocean floor food web and ecosystem. The regime shift of the late 1970s may

THIS PACIFIC SLEEPER SHARK WAS CAUGHT IN A BOTTOM TRAWL IN THE EASTERN PACIFIC OCEAN.
PHOTO COURTESY OF ELLIOTT HAZEN.

PACIFIC SLEEPER SHARKS ON THE TRAWL VESSEL DECK.
PHOTO COURTESY OF ELLIOTT HAZEN.

PACIFIC SLEEPER SHARKS BEING BRAILED FROM THE TRAWL NET.
PHOTOS COURTESY OF ELLIOTT HAZEN.

have been a result of the processes described here. As global warming changes wind patterns and the strength of the wind, we can expect more regime shifts and ones of greater magnitude (see also: "Regime Shifts and the Northeast Pacific" in the Introduction).

Another theory of reduced marine mammal populations is that great white sharks have become abundant in Alaskan waters. Sharks are secretive by nature and do not readily reveal their presence, making it difficult for scientists to study.

Sleeper shark populations are at record highs in the North Pacific, and they are preying on many species of fish and on some marine mammals. Sharks may be exerting an influence on the North Pacific marine ecosystem that will be long-lasting. Some scientists and many fishermen are concerned about what will happen, now that sharks are so common in the North Pacific Ocean. Many people have proposed shark predator-control programs without understanding the consequences.

Mathematical ecosystem models predict that there may be worse consequences if people reduce the shark population than there will be if they do not. Though sharks compete for some of the fish people catch and eat, sharks also reduce large changes in prey population numbers. According to some population models, removing sharks likely will result in increased salmon, black cod and pollock numbers. The increase in these smaller predatory fish would increase predation on smaller but extremely important forage fish such as herring, capelin and sandlance. The predicted outcome of the subsequent declines of the forage fish is for further reductions of seal and sea lion populations. This would be bad for fishermen. It also would be a big concern for those people trying to bring the sea lion back from the brink of extinction.

PACIFIC HALIBUT

Pacific halibut (*Hippoglossus stenolepis*) are in the Pleuronectidae family. Pacific halibut are the largest of the flat fish.

SIZE: Male Pacific halibut reach a maximum length of about five feet (1.5 m) while some females reach lengths over nine feet (2.7 m) and weigh over 400 pounds (181 kg). The largest sport-caught halibut ever to be weighed was a female tipping the scales at 495 pounds (224.5 kg). Her mouth was large enough to swallow a fish as large around as a watermelon. Larger halibut have been landed but not weighed.

COLOR: Halibut are dark on their upper side and white

on their underside. The color on the dark side varies to blend in with the color of the ocean floor. The underside is white and blends in with the color of the light ocean surface when viewed from below. This camouflage helps halibut to avoid detection by their prey and predators.

SPEED: Pacific halibut can swim at speeds of four to six miles per hour (6–13 km/hr).

LONGEVITY: Halibut growth rates depend upon the quality and quantity of food they obtain and the quality of their habitat. Females grow faster than males and live longer. Their ages can be determined by counting the rings on a bone in their ear, the otolith. Using this technique, scientists have determined that females can live as long as 42 years and males as long as 27 years in the wild.

REPRODUCTION: Halibut spawn during the late fall to early winter, usually at depths of 900–1,800 feet (274–549 m). Adult females in the 125–150-pound (57–68 kg) class lay approximately 3,000,000 eggs. The eggs are planktonic, drifting with the winds and currents. They hatch after about 15 days. The larvae will live on or just below the surface of the water, the top of euphotic zone, where they will drift for up to six months, possibly traveling several hundred miles (>500 km). During the free-floating planktonic stage, young halibut look much like ordinary fish, but, if they survive, their left eye will move to the right side of their head and they will take on the appearance of a flatfish. If the young halibut are fortunate, they will drift to shallow water where they begin life on the bottom. The young halibut continue to grow and develop in bays and protected waters. Most halibut spend several years rearing in shallow, near-shore waters. Males become sexually mature at age seven to eight years and females at age eight to 12 years.

SOCIAL STRUCTURE: Little is known about halibut social behavior. They sometimes group at rich feeding areas where they tend to segregate by size.

DISTRIBUTION: Pacific halibut occur throughout the North Pacific from Southern California to the Bering Sea. They occur in depths from very shallow water down to 1,800 feet (549 m).

THE UPPER SIDE OF HALIBUT COLOR CAN RANGE FROM GREENISH TO NEARLY BLACK WHILE THE UNDERSIDE IS WHITE.
PHOTOS COURTESY OF ALEXEI ORLOV.

MOVEMENTS AND MIGRATION: Fisheries biologists once believed that all halibut moved annually perhaps hundreds of miles (> 500 km) to the edge of the continental shelf to spawn and lay their eggs. Recent research suggests that many adult halibut establish and maintain

PACIFIC HALIBUT.
PHOTO COURTESY OF THE INTERNATIONAL PACIFIC HALIBUT COMMISSION.

feeding territories and may not move long distances. Young halibut, up to about 10 years of age, may be highly migratory and move in a clockwise direction throughout the Gulf of Alaska. After about 10 years, halibut become less migratory, with a smaller home range.

HABITAT: Young, free-floating (planktonic) halibut use the upper surface of the water, and juveniles prefer to live in near-shore waters. Adult Pacific halibut use a wide variety of habitat types on the ocean floor, from flat, sandy areas to rocky bottom and from shallow waters to very deep waters. They often lay in ambush on the down-current side of mounts or rock outcroppings. Adult halibut often will swim off the bottom in search of prey and even hunt just below the water's surface.

PREY: Halibut eat a large variety of animals, including salmon, cod, pollock, turbot, smaller halibut, sandlance, herring, capelin, shrimp, squid, octopus and crab. The stomachs of Pacific halibut may contain squid and octopus beaks, fish eyes and bones, hooks, rocks and wood.

PREDATORY CHARACTERISTICS: Halibut have large eyes and good vision even in the dark of the ocean depths. They often lie on the bottom to ambush their prey. They are camouflaged with their white side down and their top camouflaged with splotches of grays, black and greens. Halibut have a lateral line that is sensitive to pressure changes. This helps halibut "feel" other fish (prey) as they swim near and is helpful in the ambush. They have a large mouth with sharp teeth on top and bottom to seize their prey. Prey is swallowed whole.

PREDATORS: During their development, halibut have many predators, and adults must be wary or risk being eaten by marine mammals (sea lions and orcas) or sharks. In addition to the suite of natural predators, people also harvest halibut for food.

CURRENT STATUS: Pacific halibut populations are managed by the International Pacific Halibut Commission. Overall, the halibut numbers are stable. However, iso-

Marine Predators   89

PACIFIC HALIBUT.
PHOTO COURTESY OF THE INTERNATIONAL PACIFIC HALIBUT COMMISSION.

BRUCE STANDING ALONGSIDE A 300-POUND, 7-FOOT PACIFIC HALIBUT.
PHOTO COURTESY OF BRUCE WRIGHT.

PACIFIC HALIBUT.
PHOTO COURTESY OF JEFF CHRISTIANSEN.

PACIFIC HALIBUT.
PHOTO COURTESY OF THE INTERNATIONAL PACIFIC HALIBUT COMMISSION.

lated populations are vulnerable to over-exploitation and local depletions. This is apparent near areas where lots of sport fishing takes place. The average size of halibut being caught has declined in most areas of the North Pacific Ocean, including the Bering Sea. This indicates that the commercial fishery is having an overall negative effect on the Pacific halibut population.

ECOLOGY/CONSERVATION: The North Pacific experienced warming temperatures beginning in the late 1970s. This may have been a global warming signal. Following the warming trend the species composition in the region changed dramatically. Where fishermen had caught tons of shrimp and crab, by the early 1980s there were few of these crustaceans to be found. In contrast, numbers of pollock, cod, and some flatfish, including halibut, increased. This species composition change is referred to as a "biological regime shift." The increased abundance of halibut likely affects other species on the halibut's menu. Halibut fishermen, both sport and commercial, have experienced good catches in many areas since the early 1980s. Other halibut fishermen are not doing so well likely because of local depletions.

The International Pacific Halibut Commission (IPHC) uses a management philosophy referred to as a "precautionary approach" to managing halibut. Understanding the complexities of the marine environment is difficult at best. Most of what happens in the ocean and to marine life goes undetected by humans. This makes predicting fish population changes difficult and fisheries management a challenging science. To help avoid over-harvesting halibut, the IPHC sets very conservative harvest levels, hoping to err on the safe side of fisheries management. Some independent, non-governmental organizations would disagree and contend that the management strategy should embrace an ecosystem approach and be managed even more conservatively.

A strategy to help preserve the habitat necessary for successful reproduction and rearing of halibut and other fish and marine life is to establish marine protected areas and marine reserves. These no-fishing zones are becoming popular in some regions of the world, but there are few in the North Pacific. In the areas where marine protected areas have been established, the number, diversity and size of fish have increased. Additionally, the waters adjacent to the marine protected areas also have seen increasing fish populations, resulting in increased fish harvests overall.

Pacific halibut, being long-lived, may accumulate contaminants found in the marine environment. One of the contaminants that accumulates in halibut is mercury which is toxic to humans, especially to young children. Although mercury occurs naturally in the environment, it is a byproduct of, among other things, coal-fired power plants. Once in the environment, mercury enters waterways and accumulates in the muscle tissue of fish. The older (larger) halibut are more likely to have increased levels of contaminants, and some scientists and health officials recommend against eating these older fish.

SEA OTTER

Sea otters (*Enhydra lutris*) are members of the weasel family, which includes weasels, mink, marten, river otter and wolverine. The sea otter is the smallest marine mammal in Alaska. The sea otter captures the hearts of most who see these cute, furry creatures. They are, however, such an effective and voracious predator that they sometimes restructure entire ecosystems and are considered a keystone predator.

SIZE: Adult male sea otters can weigh 100 pounds (45.4 kg); females weigh up to 50 pounds (22.7 kg) and are about 4½ feet (1.4 m) long.

COLOR: Sea otter fur is black to pale brown with a silvery shine. Sea otter fur is the densest of any animal, with 1,000,000 hairs per square inch (155,000 hairs per cm^2). The fur insulates the otters by trapping air between the hairs, creating an insulated barrier to the cold waters they inhabit. Otters keep their fur clean and functional by meticulously grooming and cleaning. As they age, the fur on their heads turns silvery and their whiskers grow long and thick.

Sea otter.
Photo courtesy of the US Fish and Wildlife Service.

Sea otter.
Photo courtesy of the US Fish and Wildlife Service.

Sea otters.
Photo courtesy of Jim Bodkin.

ized egg may not implant immediately into the uterus but may be suspended for up to 90 days, referred to as delayed implantation. The gestation period is seven months long. Pups are usually born in the late spring. Normally, only one pup is born during each breeding cycle. At birth the pup weighs three to five pounds (1.4–2.3 kg) and is covered with a light brown, very buoyant coat of fur. The female seldom leaves the pup except to feed. When the female is on the surface, the pup often rides on her chest as she floats on her back. When she dives to feed she wraps the pup in floating kelp to prevent it from floating off and to offer some protection from predation. Bald eagles like to take unattended sea otter pups. The female sea otter produces one pup every one to two years.

SPEED: Sea otters are slow swimmers, reaching speeds of three miles per hour (4.8 km/hr) during dives. They often swim on their backs on the water's surface, again at slow speeds.

LONGEVITY: Female sea otters may live 20 years and the males 15 years.

REPRODUCTION: Females can be reproductive from ages three to 15 years. Males may be able to reproduce at an earlier age, but aggressive adult males keep the younger males from mating. Females can mate with several males in a single estrous period. The mating period lasts several days but occurs only once each year. The fertil-

SOCIAL STRUCTURE: Sea otters are usually solitary, although they may form groups, or "rafts," during the winter; non-breeding males may form bachelor groups. The bond between female sea otters and their offspring is strong during the pup's first year.

DISTRIBUTION: The sea otter population once extended from the coasts of Mexico, California, Oregon, Washington, British Columbia and Alaska, west to Kamchatka and south to Japan. Protection and relocation efforts have resulted in the recovery of sea otters into much of the available habitat.

THIS SEA OTTER HAS BEEN TAGGED SO IT CAN BE MONITORED BY RESEARCHERS. *PHOTO COURTESY OF JIM BODKIN.*

SEA OTTER.
PHOTO COURTESY OF THE .S FISH AND WILDLIFE SERVICE.

MOVEMENTS AND MIGRATION: During the winter some sea otters may move into more protected waters where large numbers of animals may raft-up in pods of 10 to 1,000 or more. Groups of non-breeding males can occur in areas adjacent to where breeding males and females occur. Population expansions of sea otters often are preceded by the occurrence of these bachelor groups.

HABITAT: Sea otters feed in waters from five to 250 feet (1.5–76 m) deep, and they feed from close to the beach to over a mile (1.6 km) offshore.

PREY: Sea otters eat a variety of near-shore animals including sea urchins, crabs, clams, mussels, octopi, other invertebrates and fishes.

PREDATORY CHARACTERISTICS: The otter's front feet are strong and adapted for digging and handling slippery prey. The back feet are webbed, making them useful for swimming. Sea otters use their tails for propulsion. However, when diving they employ body undulations using both their tail and hind feet.

Sea otters use tools, such as a rock placed on their bellies, as they float on their backs. They break clams and other hard-shelled prey on the rock and discard the shells before eating the soft tissue. Some sea otters have special rocks they use over and over again.

An adult sea otter may consume 35–40 percent of its body weight in food each day to meet its nutritional and energy needs. Their daily food consumption increases during the cold winter or when females are nursing.

PREDATORS: Orcas and sharks will eat adult sea otters, and bald eagles may take unattended young. Alaska Native people sometimes harvest sea otters for their fur.

CURRENT STATUS: Russian fur traders reduced the sea otter population from about 150,000 animals to 2,000 animals by 1911. Protection and reintroduction efforts have resulted in the recovery of sea otters into all the general areas they once inhabited in Alaska. Their populations remain low in most areas outside of Alaska.

ECOLOGY/CONSERVATION: Few animals effect and restructure a near-shore ecosystem to the extent that otters can. This is obvious in areas recently colonized by otters. Ecosystems without sea otters are often home to a large population of sea urchins. These prickly marine grazers consume vast amounts of marine algae such as large kelps. In addition to eating the plants, sea urchins will eat through the kelp holdfasts that keep the plants from floating off and washing up on shore. When the otters move in, the sea urchin population usually gets wiped out. This allows the kelps to return and flourish. The thick kelp beds offer cover and food for many fish species that soon re-colonize the near-shore area. When sea otter populations decline, sea urchin populations return and large marine kelp beds are soon reduced.

Large natural swings in sea otter numbers have been observed in Alaska's remote Aleutian Islands. During the 1990s the Aleutian Islands' sea otter population declined by nearly 90 percent, dropping from about 10,000 animals to 1,000 animals. A small pod of marine mammal-consuming orcas, called transients, frequented the area and were seen eating otters on a couple of occasions. Based on the energy demands for orcas, only four of them could account for the decline in the Aleutian Island sea otter population. Following the disappearance of the sea otters, sea urchins are recovering and eliminating the kelp beds. The impact of only a few predators on this large, near-shore marine ecosystem is dramatic and likely will be long lasting. This type of effect is referred to as top-down control.

Sea otters also are very susceptible to the effects of oil spills and oil pollution. Acute mortalities usually result when oil fouls an otter's fur, causing loss of body heat, hypothermia and death. When oiled sea otters attempt to clean the oil from their fur, they ultimately ingest the toxic pollutant or inhale the toxic vapors, either of which can result in death. Chronic effects and death occur when otters eat animals contaminated with oil. Mussels and clams, important otter prey, accumulate and concentrate contaminants, including oil. Otters can die even years after an oil spill if they continue to eat contaminated prey.

The *Exxon Valdez* oil spill of 1989 killed thousands of sea otters within the first few months of the spill (acute mortalities). Much of the spilled oil ended up buried in the beaches. This oil continuously enters the food chain in small amounts, poisoning the food web for decades. Over 20 years after the oil spill, sea otters are still dying from *Exxon Valdez* oil.

Rehabilitation centers were set up to clean and help wildlife contaminated by Exxon's oil. The cost for capture, cleaning and rehabilitating sea otters was $80,000 each. Many of the otters were later released back into the environment. Nearly all of these otters later died because of permanent damage to the otters or later contamination from the oil that remained in the environment. To protect sea otters from oil pollution we must re-double our efforts to prevent oil spills and reduce our dependence upon this highly toxic commodity.

THIS SEA OTTER DIED AFTER BEING COATED WITH SPILLED CRUDE OIL. PHOTO COURTESY OF THE US FISH AND WILDLIFE SERVICE.

Steller's Sea Lion

The Steller's sea lion (*Eumetopias jubatus*), also known as the northern sea lion, is the largest of the eared seals. Sea lions differ from seals by having external ears and rear flippers that allow them to walk on land. The sea lion's name is said to come from their thick neck that resembles that of a lion's mane. Their loud roar also may remind one of a lion.

SIZE: Adult male Steller's sea lions average 11 feet (3.4 m) in length and weigh an impressive 1,300 pounds (590 kg). However, some males have reached weights exceeding 2,000 pounds (907 kg). They reach this size at around 12 years of age. Females reach adult size at about seven years when they are nine feet (2.7 m) long and weigh approximately 600 pounds (272 kg).

COLOR: Sea lions are tan to reddish-brown in color when dry. They are much darker when wet.

SPEED: Steller's sea lions can obtain speeds of about 20 miles per hour (32 km/hr) in the water. They are slower on land.

LONGEVITY: Both males and females may live 20 years.

REPRODUCTION: Females become sexually mature between ages three and eight years. Bulls are able to establish breeding territories and mate when they are nine to 13 years old. Sea lions use the same beaches or rookeries each year to breed and give birth. Bulls establish breeding territories from mid-May through mid-July. During this time the larger, stronger and most aggressive males establish and maintain the prime areas of the rookery for breeding. These dominant males will mate with many females. During the breeding season the dominant bulls do not eat or leave their territory unless forced out by a stronger bull.

Females give birth to a single pup from mid-May through July. The female breeds shortly after giving birth. The fertilized egg does not implant in the uterus until October. This is referred to as delayed implantation. Females give birth to a single pup that weighs about 50 pounds (23 kg) and is about 45 inches (1.1 m) long. Pups have dark brown fur with a colorless tip on the end of each hair, giving them a frosty appearance. They become lighter in color with age. Mothers nurse their pup at the rookery for about two weeks before returning to sea to feed. The mothers return often to feed their pup. The female is able to locate and feed her pup on the noisy and crowded rookery, probably by identifying the pup's unique smell or cry. Weaning age varies for pups, as some are weaned at one year while others continue to nurse for up to three years.

SOCIAL STRUCTURE: Sea lions are very social and often are seen in groups when swimming or at haul-out sites.

DISTRIBUTION: Steller's sea lions are found in the North Pacific from California and northern Japan to the Bering Sea. Their population is divided into the western stock (Gulf of Alaska, Bering Sea and the waters off Russia and Japan) and the eastern stock (California, British Columbia and Southeast Alaska).

MOVEMENTS AND MIGRATION: Steller's sea lions can travel for great distances. The longest recorded movement was of a sea lion tagged near Kodiak, Alaska, and later recorded near Ketchikan, Alaska, a distance of over 900 miles (1448 km). They move to areas with high prey abundance (hot spots) such as forage fish spawning events and salmon runs.

MALE STELLER'S SEA LION ON THE BEACH.
PHOTO COURTESY OF THE US FISH AND WILDLIFE SERVICE.

Marine Predators

HABITAT: Steller's sea lions feed in a variety of marine habitats from the intertidal zone to the continental shelf, many miles offshore. They find prey from the ocean surface to the ocean floor.

PREY: Sea lions are near the top of the marine food web and feed on a variety of prey including squid, octopus, capelin, sandlance, herring, eulachon, flounder, salmon, cod, rockfish, pollock and sculpin. Steller's sea lions have been observed eating other marine mammals, including fur seals. It might surprise some at how similar their list of diet items are to salmon sharks, which share the same waters.

PREDATORY CHARACTERISTICS: Sea lions use a variety of interesting hunting techniques. They may hunt singly or in groups of a few to hundreds of individuals. They often spend time drifting and swimming near points of land and narrow passes where fish are concentrated. When salmon are available, sea lions will hunt the migrating fish and follow schools right up their spawning streams and rivers. When it catches large fish, such as an adult salmon, the sea lion will lift its head out of the water and give a violent shake, ripping the fish to pieces that are easily swallowed.

During spawning of some small schooling forage fish (herring, capelin, eulachon and sandlance), sea lions often hunt in large groups, or "pods," with sometimes more than 50 individuals. The sea lions will swim side-by-side, herding the fish into waters where they can be caught. Small fish are just gulped down whole.

STELLER'S SEA LIONS.
PHOTO COURTESY OF THE US FISH AND WILDLIFE SERVICE.

STELLER'S SEA LIONS.
PHOTO COURTESY OF THE US FISH AND WILDLIFE SERVICE.

STELLER'S SEA LIONS AT A SOUTHEAST ALASKA HAUL OUT.
PHOTO BY BRUCE WRIGHT.

STELLER'S SEA LION.
PHOTO COURTESY OF THE U.S. FISH AND WILDLIFE SERVICE.

STELLER'S SEA LIONS.
PHOTO COURTESY OF THE U.S. FISH AND WILDLIFE SERVICE.

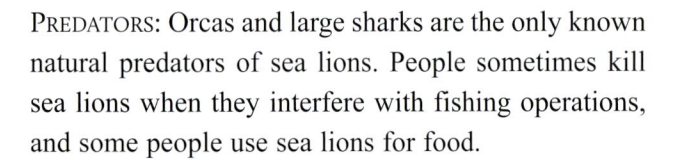

PREDATORS: Orcas and large sharks are the only known natural predators of sea lions. People sometimes kill sea lions when they interfere with fishing operations, and some people use sea lions for food.

CURRENT STATUS: Alaska's Steller's sea lion population was 242,000 in the 1970s, but Alaska's western stock declined by more than 50 percent by the mid-1980s and is down by 90 percent in some areas. The western stock is listed as endangered under the Endangered Species Act.

ECOLOGY/CONSERVATION: Forage fish (herring, capelin, sandlance and smelt) are high in the essential nutrients and fatty acids important for marine predators. An interesting change occurred in the North Pacific Ocean in the late 1970s. The region's waters once were teaming with lots of forage fish, particularly capelin. During that period the seabird and sea lion numbers and productivity were both very high. When the forage fish population declined in the late 1970s, sea lion and some seabird populations declined as well. Other North Pacific marine populations increased, including arrowtooth flounder, salmon, cod, pollock and sharks. This region-wide change in the species composition is referred to as a "biological regime shift."

The sea lion population crash has led to the western stock being listed as endangered under the Endangered Species Act. Although scientists are not sure what

THIS STELLER'S SEA LION IS PHOTOGRAPHED FROM A RESEARCH SUBMERSIBLE. PHOTO COURTESY OF DAVID SCHEEL.

caused the decline in sea lions, some of the causes being considered for declines in predator populations are: disease, effects of contaminants, predation, loss of critical habitat, or loss or change of important food supplies. Sea lions in the North Pacific carry contaminant loads, are eaten by orcas and sharks, are shot by fishermen, and experienced a dramatic decline in their most important food base, forage fish. Any or all of these factors likely contributed to the decline of sea lions.

The dramatic increase of other fish predators, especially arrowtooth flounder and pollock, may have reduced the forage fish numbers and may keep the critical forage fish populations depressed at a small fraction of what they were before the regime shift. You may ask, where did all the flounder and pollock come

STELLER'S SEA LIONS.
PHOTO COURTESY OF THE U.S. FISH AND WILDLIFE SERVICE.

from, and was there a connection with the regime shift? Scientists, who are trying to figure out ways to keep the sea lions from going extinct, are asking that question. You have learned from reading the section on sharks that some predator populations actually may moderate dramatic changes in the ecosystem and help to stabilize the system. Some fishermen and scientists have proposed catching most of the pollock and flounder, hoping the forage fish population may respond and increase. However, ecosystems are very complex, and reducing one predator population could disrupt the balance and promote another predator population. There are no simple explanations for understanding the changes in marine ecosystems and there are no easy ways to manipulate marine ecosystems to suit the needs of people.

The loss of a sea lion population in the North Pacific ecosystem would be a terrible thing. They once were so abundant that they were used as food by the Native peoples of Alaska; today they rarely are used for food.

Each species we lose on Earth would be analogous to a bridge losing rivets, one at a time. You would not notice much at first, and then the bridge may groan in the wind and creak and crack. Finally, when enough rivets have been lost, the bridge will crash and all is lost. Many scientists are warning us that the bridge is groaning; the loss of rivets (species) must stop before it is too late. We cannot afford the extinction of Steller's sea lions or any other species.

Polar Bear

The polar bear's scientific name, *Ursus maritimus*, means sea bear. Polar bears are the most carnivorous of the bear family. Alaska has two recognized polar bear populations: the Beaufort Sea and Chukchi/Bering Sea populations.

SIZE: Polar bears are large; males can weigh 1,700 pounds (771 kg) and be eight to 10 feet (2.4–3 m) long. Females reach a weight of 700 pounds (318 kg) and six to eight feet (1.8–2.4 m) in length. Compared to the closely related brown bear, polar bears can weigh more, but they have a smaller head and longer neck. Their body is stocky, but they lack the large shoulder hump seen on brown bears.

COLOR: Polar bear colors range from pure white to yellowish to gray or brownish. The darker colors can be due to staining by seal blubber. The nose, skin and lips are black. Their pelage, or fur, consists of a thick undercoat and an abundance of long guard hairs.

SPEED: Polar bears can swim six miles per hour (9.7 km/hr) and can run 25 miles per hour (40 km/hr).

LONGEVITY: Polar bears can live at least 32 years in the wild.

REPRODUCTION: From late March through May male polar bears seek out females. They mate with as many as they can during the spring mating season. Mating usually occurs out on the sea ice. The fertilized egg does not implant until September, and it may not implant and develop if the female is unhealthy. This is called "delayed implantation."

The female locates a suitable den site on either sea ice or land. The den is excavated in the snow, usually under a snow bank or on uplifted sea ice. An enlarged den chamber is used to give birth to the cubs (usually two), which occurs in December. Unlike black and brown bears, polar bears do not hibernate. The sole purpose of the den is to protect the bear family when the cubs are very young. The newborns weigh about a pound (0.5 kg) at birth, are blind and covered with fur. They grow quickly on rich bear milk, which has a 30 percent fat content. The bears emerge from the den in the spring, late March or early April, after the cubs reach about 15 pounds (6.8 kg). They will continue to use the den until the cubs become acclimated to the outside environment; then they begin traveling on the drifting sea ice. The mother bear and her young stay together for about 2½ years before they separate and she breeds again.

Females do not breed until they are at least four years old and sometimes not until they are eight years old. They can produce young every three years, or about five litters during their lifetime, a relatively low reproductive rate, but similar to that of brown bears.

SOCIAL STRUCTURE: Polar bears are usually solitary, except females with young, groups feeding on an abundant food patch such as a beached whale, or during the breeding season.

DISTRIBUTION: Polar bears can occur throughout the North Polar Region. They are most abundant near coastlines and the southern edge of the pack ice.

MOVEMENTS AND MIGRATION: The edge of the ice moves seasonally, and polar bear movements tend to follow the shifting ice south during the winter and north during the summer. In Alaska the southern extent of polar bears is St. Matthew Island and the Kuskokwim River delta in the Bering Sea. The extent of a polar bear's annual movement, or its home range, is determined by food availability, weather, and ice and ocean conditions. Some polar bears may travel thousands of miles (>5,000 km) each year in search of food, mates and den sites.

HABITAT: Polar bears use the ice and swim in the ocean in their search for prey. When the ice melts in summer they may be left stranded on land for months at a time.

PREY: Polar bears are primarily carnivores eating ringed seals, bearded seals, walruses, beluga whales, small mammals, birds and bird eggs.

Marine Predators 99

POLAR BEAR CUBS.
PHOTO COURTESY OF THE US FISH AND WILDLIFE SERVICE.

FEMALE POLAR BEAR WITH HER CUBS.
PHOTO COURTESY OF THE US FISH AND WILDLIFE SERVICE.

POLAR BEAR.
PHOTO COURTESY OF JOHN GOMES AND THE ALASKA ZOO.

POLAR BEAR.
PHOTO COURTESY OF JOHN GOMES AND THE ALASKA ZOO.

POLAR BEAR HUNTING ON THE ICE.
PHOTO COURTESY OF THE US FISH AND WILDLIFE SERVICE.

POLAR BEAR FEEDING ON A SEAL.
PHOTO COURTESY OF THE US FISH AND WILDLIFE SERVICE.

THIS POLAR BEAR JUMPS ACROSS AN OPENING, BUT GETS WET, THEN ROLLS IN SNOW AND SHAKES TO DRY ITSELF.
PHOTOS COURTESY OF JOHN GOMES AND THE ALASKA ZOO.

Marine Predators   101

They also will eat carrion, including large whales washed up on beaches. Polar bears rarely attack and kill people. Since 1970 polar bears have killed seven people in Canada's arctic region, and one person in Alaska.

PREDATORY CHARACTERISTICS: Polar bears' paws are adapted to their harsh environment. The paws of an adult can measure up to 12 inches (30 cm) across, allowing the bear to travel on thin ice. The black footpads are covered with bumps, and their sharp, 2-inch (5.1 cm) curved claws are adapted for traveling on the ice and securing prey. The large forepaws are used to paddle through the water, and the rear paws serve as steerage. Hair covers the bottoms of their feet.

The senses of smell and vision of the polar bear are very keen, probably superior to that of the brown bear. A polar bear's teeth are longer, sharper and more widely spaced than those of the brown bear, adaptations suited to the polar bear's diet of seals, walruses and small whales. Another adaptation that polar bears have is a layer of blubber over four inches (10.2 cm) thick which acts as insulation to keep polar bears warm, even when they are swimming in ice-filled waters. They are so well insulated that they must be careful not to overheat.

Polar bears spend much of their time hunting their favored prey, the four-foot (1.2 m) long, 150-pound (68 kg) ringed seal. The bears prize the ringed seal's thick layer of blubber, used to insulate the seal from the bitterly cold waters. Polar bears lie in wait near active seal breathing holes and grab the seals when they come up to breathe. This is the technique used most often during the winter when the sea is mostly frozen over. During early summer, ringed seals bask in the warm sun. Their dark coloration can be spotted a long way off by the alert bear. The bear slowly and quietly stalks the seal, then, when about 20 feet (6.1 m) away, the bear charges, capturing the seal before it can escape.

In spring, adult seals will leave their very young pups in sheltered snow chambers on the ice near their breathing holes, so the female can dive for food. Polar bears will seek out these unsubstantial shelters and prey on the pups. When prey is plentiful polar bears will eat only the energy-rich blubber and skin, leaving the remains for younger inexperienced bears, arctic foxes and other scavengers.

THESE POLAR BEARS ARE CHECKING OUT THE SUBMARINE THAT SURFACED NEAR THE NORTH POLE.
PHOTOS COURTESY OF THE US NAVY.

Polar bears survive because they can manage through a feast-and-fasting lifestyle. When food is abundant they must take advantage of the situation. A beached whale carcass can attract dozens of polar bears. One bear may dominate, or own, the carcass, but if a specific polar bear protocol is followed, all incoming polar bears will be allowed to feast on the whale. The trespassing bears beg in a submissive manner with a low-to-the-ground, slowly circling approach. This is followed by nose touching with the bear that owns the whale.

Polar bears will kill and eat beluga whales trapped in a small opening in a vast expanse of ice. Dozens of whales may be taken in this manner from a single, small opening.

After eating, a polar bear is likely to be soiled with blood and blubber. They are careful to clean themselves, either in water or, if water is not available, in snow. A polar bear will carefully lick its chest, muzzle and paws. Females will lick their cubs to keep them clean and teach them how to wash in water and snow. Polar bears are careful to keep their fur clean, because soiled fur has reduced insulation properties and because their ability to blend in with the background of white also is reduced.

PREDATORS: Polar bears may be killed by other, larger polar bears. Their only other natural predator is the orca. People hunt polar bears for their hides and as trophies. Some people eat polar bear meat.

POLAR BEAR.
PHOTO COURTESY OF THE US FISH AND WILDLIFE SERVICE.

CURRENT STATUS: Polar bear populations have increased slowly in Alaska since the 1970s. The Alaska population of polar bears is estimated at 3,000–5,000 animals, and the world population of polar bears is estimated at 22,000–27,000 animals.

ECOLOGY/CONSERVATION: Under State of Alaska management, polar bear populations were decreasing and at risk. Alaska allowed hunting of polar bears from the ground and from aircraft, including from helicopters. With the passage of the Marine Mammal Protection

POLAR BEAR.
PHOTO COURTESY OF JOHN GOMES AND THE ALASKA ZOO.

Act in 1972, management authority of polar bears was transferred to the federal government. A moratorium was placed on hunting polar bears, except by Alaska Natives, and use of aircraft was prohibited. The hides and meat of polar bears are used by Alaska Natives, and the tanned hides are used in traditional arts.

Still, the greatest threats to the survival of the polar bear now are due to human activities. Oil and gas development in the Arctic poses an increasing risk to bears. Oil spills in the Arctic are difficult or impossible to clean up. If the fur of a polar bear is contaminated with oil, the insulation value of the fur is reduced, stressing the animal. Ingesting oil by cleaning its fur or eating oil-contaminated prey and breathing toxic petroleum fumes could kill polar bears, affecting polar bear populations over a very large area.

Contaminant levels, particularly PCBs and DDT, continue to increase in the Arctic. These toxic compounds concentrate as they move up the food web in a process called "biomagnification." These contaminants have reached high levels in many Alaskan predators. Scientists are working to predict how these toxins ultimately will affect Alaska's predators, ecosystems and people.

Global warming ultimately may exert the greatest impact on the Arctic. The size and thickness of the Arctic ice cap has diminished, altering polar bear habitat. During the summer many polar bears have difficulties securing food; some bears are stranded on land waiting for the ocean to freeze so that they can continue to hunt seal. Recent observations of drowned polar bears may be a harbinger of things to come for this species. Many polar bears are shot each year when they find their way into villages and interact with the human residents. Polar bears depend upon the ice flows and cold temperatures, an environment for which they are adapted and are well suited.

Orca

The orca (*Orcinus orca*), also known as killer whale, is the largest member of the dolphin family, in the suborder, odontoceti, or "toothed whales."

SIZE: Relative to the great whales, orcas are small. The males, or bulls, are 19–32 feet (5.8–9.8 m) long and weigh 8,000 to 22,000 pounds (3600–10,000 kg). The females, or cows, are 16–28 feet (4.9–8.5 m) long and weigh 3,000–16,000 pounds (1360–7300 kg). The tall, black dorsal fin is characteristic of this species. The male's dorsal fin may reach six feet (2 m) in height, and the female's may reach three feet (1 m).

COLOR: Orcas are recognized by their striking black and white coloration. They are all black except for the area below and behind the dorsal fins, the bottom surface of the lower jaw, the undersides of the tail flukes and the eye patches slightly behind each eye, which are white. The area just behind the dorsal fin is gray and is called the "saddle patch." The saddle patch is photographed by researchers and used to identify individual animals. The black-and-white coloration and patterns may act as camouflage in the ocean in which

ORCA. *Photo by John Harvey.*

ORCA. *Photo by John Harvey.*

ORCA POD. *PHOTO COURTESY OF JOHN HARVEY.*

ORCAS SPY HOPPING FROM A BREATHING HOLE IN THE ICE. *PHOTO COURTESY OF NATIONAL OCEANIC AND ATMOSPHERIC ADMINISTRATION.*

the disruptive coloration contradicts the animals' size, shape, and direction of movement.

SPEED: Orcas are one of the fastest marine mammals, reaching speeds of 25 miles per hour (40 km/hr), but they generally cruise at two to six miles per hour (5–10 km/hr). The great bursts of speed require great strength and lots of energy. Orcas are mammals, so they must breathe air (oxygen) from their blowhole, located on the top of their heads. They usually hold their breath for four to five minutes before surfacing and breathing several times, but they can hold their breath for up to 15 minutes and can dive down 900 feet (275 m). There is no evidence that orcas suffer from the bends, nitrogen bubbles in the blood that can cause death in humans.

LONGEVITY: If they survive the first year, male orcas may live 35 years or more and females may live 50 years or more in the wild.

REPRODUCTION: Females reach reproductive age at about 12–15 years old (15–16 ft or 4.6–4.9 m), and males when they are 12–15 years old (18–20 ft or 5.5–6.1 m). Orcas are thought to be polygamous and mate with a number of partners. Breeding can occur at any season, but in Alaska it usually occurs during the summer. The gestation period is the longest of all marine mammals, about 17 months. The calf can be born headfirst or tail first. Under good conditions, a female may give birth every three years, but sometimes as many as 10 years pass between calving years. Twins are very rare.

Calves are about eight feet (2.4 m) long at birth and weigh 300–400 pounds (136–181 kg). All orca whales nurse their young, and calves nurse underwater. The milk is high in fat (30–40 percent), and the calf may nurse for more than 12 months. Calves grow about four inches (10.2 cm) per month during their first year. Calf mortality in the wild can be as high as 50 percent.

SOCIAL STRUCTURE: Orcas live in social groups, usually of related individuals, called "pods." In the waters around Alaska, pods are generally five to 30 animals. Within the pods are sub-pods, or maternal groups, containing females and their offspring. The males usually swim off to the sides at some distance. During breeding there will be temporary exchanges between members of different pods. At these times, super-pods of more than 100 animals have been observed.

In Alaska's waters orcas have been categorized into three group types: residents, transients and offshore whales. The resident whales do not leave the pod for extended periods of time, are more vocal, have larger pods (five to 50 animals) and eat mostly or only

THIS ORCA IS SPY HOPPING THROUGH A HOLE IN THE ICE. PHOTO COURTESY OF NATIONAL OCEANIC AND ATMOSPHERIC ADMINISTRATION.

fish. The transients have a more flexible social structure in that animals may leave the pod for many years, are less vocal and more secretive, travel in small groups, usually two to seven individuals, and eat only or mostly other marine mammals. Little research has been done on the offshore type, so scientists have yet to determine their social attributes.

Orcas communicate by body movements and displays, biting, slapping their tails, and producing clicks, whistles, moans, squeaks and other sounds. Different groups have different dialects, and the greater the geographic distances between groups the more likely the dialects are to differ.

DISTRIBUTION: Orcas can be found in all the oceans of the world.

MOVEMENTS AND MIGRATION: Orca movements are often in response to prey availability such as herring and salmon spawning events, and baleen whale migrations. In the northern and southern oceans orcas migrate in response to the advance and retreat of the pack ice. Pods can travel thousands of miles a year and have a home range of tens of thousands of square miles.

HABITAT: Orcas are most common in areas of high ocean productivity and strong ocean upwellings. These upwelling areas, or hot spots, regularly have greater concentrations of prey. Orcas have been seen as far as 100 miles (161 km) up rivers in pursuit of prey. Orcas are less common in tropical waters.

PREY: The orca is at the top of the marine food web. Their diet items include fish, squid, seals, sea lions,

ORCAS.
PHOTO COURTESY OF THE US FISH AND WILDLIFE SERVICE.

ORCA.
PHOTO COURTESY OF THE US FISH AND WILDLIFE SERVICE.

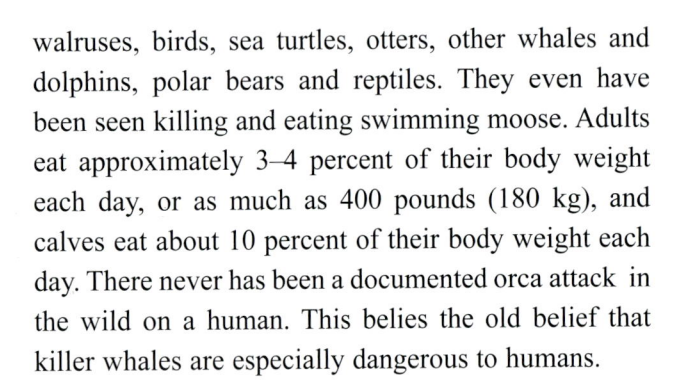

walruses, birds, sea turtles, otters, other whales and dolphins, polar bears and reptiles. They even have been seen killing and eating swimming moose. Adults eat approximately 3–4 percent of their body weight each day, or as much as 400 pounds (180 kg), and calves eat about 10 percent of their body weight each day. There never has been a documented orca attack in the wild on a human. This belies the old belief that killer whales are especially dangerous to humans.

PREDATORY CHARACTERISTICS: Orca teeth are conical and interlocking, useful for grabbing fish or ripping flesh from larger animals. They have 10–14 teeth in each side of each jaw, 40–56 teeth in all. Each tooth is about three inches (8 cm) long and one inch (2.5 cm) wide, coming to a sharp point at the end. Orcas have keen eyesight, both in and out of the water. They use high-frequency clicks and subsequent echoes to locate objects and prey in the water, which happens swiftly. Due to water's greater density, sound travels four and a half times faster in water than in air.

Feeding habits of two types of orcas (resident and transient orcas) have been described. Resident orcas feed primarily or exclusively on fish, and transient orcas feed primarily or exclusively on mammals, usually other marine mammals. The social order and hunting strategies are reflected in the prey type that these two groups of orcas eat. The residents often hunt in large spread-out groups, using echolocation to locate prey and to communicate with other pod members, possibly to coordinate the attack. They hunt open waters and channels from shore to shore. Transients are stealthier, sneaking along close to the shoreline in silence. Their blows in breathing are less obvious, and they may not vocalize at all until after the kill. Transients dive and swim for a mile (1.6 km) or more underwater to come up right in front of sea lion haulouts, to catch the animals unaware.

Orcas hunt cooperatively, much as do wolves or African hunting dogs. They can surround their prey and attack from different directions. Once their prey is surrounded, the whales can take their time to pick off one or more of the target animals. Seals, sea lions, and porpoises may turn into objects of what appears to us to be play. The orcas may toss the prey high into the air, batting it about. Orcas attack and kill great whales, including the largest animal ever to have occurred on earth, the blue whale. Entire pods of the largest toothed whale, sperm whales, great predators themselves, have been attacked and killed by orcas.

Orcas attack and eat the most dangerous fish, including white sharks and salmon sharks. The whales use their stealth and speed to get the jump on these prey items, often striking the sharks in their mid-section from below. Orcas have been observed leaping from the water with 1,000-pound (450 kg) sharks, or 2000-pound (900 kg) sea lions in their jaws. Orcas may attack prey on land, such as sea lions, or prey that is on ice floes, such as walruses and polar bears. They have been observed rolling icebergs from below, causing prey to slip into the water where it can be killed and eaten. When taking on large prey they will swallow great chunks of meat, an easy task due to their large throats. An adult orca can swallow seals whole.

Resident orcas appear to tolerate the presence of marine mammals. Dall's porpoise have been observed at "play" with members of a resident orca pod. The play can last for hours. The porpoise can detect that resident whales are not a threat.

PREDATORS: Adult orcas have no natural predators, but large sharks may take the calves. Humans are the only significant predators of orcas. People have taken orcas for their meat and oil, for entertainment and research purposes at marine parks, and have shot and killed them when they interact with fishing activities.

CURRENT STATUS: The orca populations appear to be doing well in Alaska, but there is concern that high levels of contaminants may affect them, reducing their numbers in the near future. Alaska's waters harbor about 250 orcas (15 pods from Prince William Sound to Kodiak Island), and approximately 160 animals have been identified in Southeast Alaska waters.

Another 100 have been identified along the Aleutian Islands and another 100 in the Bering Sea. It is likely that many orcas in Alaska have not been photo-identified and remain uncounted.

ECOLOGY/CONSERVATION: Orcas can have a profound impact on the marine food web. Thousands of sea otters went missing from the Aleutian Islands during a ten-year period, 1990–2000. Some scientists believe that the 40,000 missing sea otters could have been removed (eaten) by as few as four orcas.

These changes may be a result of the harvest of the great whales during the whaling era. The hypothesis predicts that once the whalers killed the great whales, the transient orcas switched prey to seals, sea lions and sea otters, a phenomenon referred to as a "trophic cascade." This hypothesis is disputed by other scientists, and additional research needs to be done in order to understand the effects of humans and other top predators on the marine ecosystem.

ORCA. PHOTO BY JOHN HARVEY.

ORCA. PHOTO COURTESY OF THE US FISH AND WILDLIFE SERVICE.

Now that the sea otter population is a fraction of what it once was, the sea urchin population, a favored sea otter food, is booming, eating the marine kelps and altering the near-shore ecosystem. This is referred to as top-down control and is an important factor in altering marine ecosystems.

Orcas are at the top of the marine food web, so they are more susceptible to contaminants than are many other species, due to biomagnification and the increase of contaminants in the marine ecosystem.

Bruce Wright stands alongside a young male transient orca that died and was beached near Cordova, Alaska. A month earlier Bruce watched this whale feeding in Prince William Sound and wondered if its diet of marine mammals might be putting it at risk of disease or death due to contamination. Biopsies taken from the dead whale later showed high levels of persistent organic pollutants including PCBs and DDT. *Photo courtesy of Bruce Wright.*

Biomagnification causes the concentrations to become so high that they impair the orcas' immune and reproductive systems or kill orca offspring.

Orcas and other marine mammals can be affected in the vicinity of oil spills. In 1989 two pods of orcas were impaired by the *Exxon Valdez* oil spill in Prince William Sound, Alaska. After more than 20 years, one of the two pods, the transient AT pod, have yet to recover from that catastrophe.

Whale watching has become a popular tourist attraction. Ecotourism, once thought to be a non-impact industry, has grown to the point that wildlife and the environment are being assaulted. Some whale-watching boats approach and disturb orcas and other whales, causing stress and keeping the whales from feeding. Federal regulations have been enacted to help reduce the effect of this industry. These laws need to be enforced to provide whales with a buffer so that they can feed and interact with each other.

Great White Shark

Great white sharks (*Carcharodon carcharias*) — also known as great white, white pointer, white shark, or white death — are fish, and are in the Lamnidae family of sharks. This family includes the salmon and mako sharks. Lamnidae sharks are warm-blooded (partially endothermic) and intelligent. Great white sharks are one of the most notorious predators.

SIZE: Great white sharks can reach and likely exceed 21 feet in length (6.4 meters) and weigh 7,330 pounds (3,224 kg). The larger great white sharks may occur in colder waters such as the food-rich cold waters of the North Pacific and Arctic Oceans.

COLOR: Like salmon sharks, great white sharks are dark gray above and the underside is whitish. The shark's colorations help camouflage it both from above and below, an important advantage for this predator as it searches for and attacks unsuspecting prey.

SPEED: Great white sharks can swim at speeds approaching 25 miles per hour (40 km per hour) with burst speeds to 35 miles per hour (56 km per hour).

LONGEVITY: Great white sharks reach maturity around 15 years of age and have a life span of over 30 years. Most fish are aged using bony structures called otoliths; however, sharks do not have bones, making it difficult to age them.

REPRODUCTION: Mating requires the male shark to secure the female by grabbing her with his mouth and inserting his clasper organs into the female for copulation. Females may have bite marks along their flanks and on their pectoral fins indicating they had recently mated. White shark mating is apparently not a gentle affair, but this may be further evidence that great white sharks have a limited sense of pain.

WHITE SHARK.
PHOTO COURTESY OF R. AIDAN MARTIN.

SOCIAL STRUCTURE: Observing and measuring social behavior of great white sharks is difficult as they don't allow for easy observation especially in the often cloudy waters off Alaska. Still, evidence is mounting that indicates sharks are socially complex and benefit from these qualities. Feeding hierarchies may be established at locations with abundant prey such as seal and sealion rookeries and floating whale carcasses. An observed attack on beluga whales in

Cook Inlet indicates what appeared to be a complex, cooperative attack. This is of special interest because the opaque water reduced or eliminated visual cues during the attack.

DISTRIBUTION: Great white sharks live in almost all coastal and offshore waters of the world. It was once thought that white sharks were only found in warmer waters with temperatures between 54 and 75°F (12 and 24°C), but observations of white sharks in Alaska waters with temperatures approaching freezing indicates they can use sub-arctic and arctic waters too.

MOVEMENTS AND MIGRATION: Satellite tags attached to great white shark dorsal fins have revealed that they are highly migratory, just like salmon sharks, and migrate between Baja California and Hawaii, between Australia and South Africa and between South Africa and the Indian Ocean. The migration swimming speeds appear to be steady and the distance covered annually can exceed 12,000 miles (20,000 km).

When the Baja–Hawaii white sharks arrive in the Hawaiian Islands their swimming behavior changes to shallower excursions, but the reasons for these migrations and differing behaviors remain a mystery. White sharks were noted using Alaska waters in the 1970s, but as more observations have been compiled they appear to use Alaska waters year round. It is not known how far north in the Bering Sea white sharks travel, but their travels are likely limited only by food availability; if they secure enough food they likely can use even Alaska's coldest marine waters.

HABITAT: Great white sharks can be found patrolling nearshore waters as they search for prey. They appear to be regular visitors to the waters of Southeast Alaska, off Yakutat in Prince William Sound, and they have been seen several times in Cook Inlet, along the Alaska Peninsula and in the Aleutian Islands. They likely also use offshore waters much like salmon sharks where they find concentrations of prey.

NOTE THE LARGE TRIANGULAR-SHAPED TEETH ON THIS WHITE SHARK. © SAVE OUR SEAS LTD. PHOTO BY TOM CAMPBELL.

PREY: Great white sharks prey upon halibut, salmon, tuna, rays, other sharks, dolphins, porpoises, whales, hair seals, fur seals, elephant seals, sea lions, sea turtles, sea otters, seabirds and invertebrates. As they become adult and get larger, great white sharks take large prey and more marine mammals. Large great white sharks have been observed taking beluga whales in Cook Inlet; they may take walruses in the Bering Sea and Arctic Ocean.

PREDATORY CHARACTERISTICS: Great white sharks are opportunistic, but, like other predators, individual white sharks likely become proficient at taking a few prey species and other prey species only irregularly. The reason for this is that the techniques for locating and safely securing large and dangerous prey are different for each species preyed upon. For example, an adult white shark that's learned to take seals may have learned to drag the adult seals to the bottom until the prey has drowned after which it is consumed. Larger, more dangerous prey, such as adult elephant seals, are more likely to be struck from behind and allowed to bleed out and die before the shark returns to feed; or, for another individual great white shark, relentless and continued attack may be preferred. A great white shark that is successful with a certain prey species may be less likely to bother learning the necessary skills and techniques needed to diversify its diet. Some great white sharks have learned to scavenge whales killed by orcas. Adult great white sharks tend to prey more on marine mammals than fish, preferring prey with high contents of energy-rich fat.

ECOLOGY/CONSERVATION: Great white sharks locate and utilize areas of high food abundance, or "hot spots." The stomach of a large dead great white shark found beached in Southeast Alaska contained undigested salmon parts, but large great white sharks tend to eat marine mammals because of their higher fat content.

Great white shark predation in Alaskan waters may afford some stability to the ecosystem by removing and controlling other large marine predator. Great white shark predation may be limiting the predation by intermediate-sized predators. Great white shark predation on beluga whales in Cook Inlet may be pushing this population of white whales to the brink of extinction. The complexities of the predator–prey relationship and their effects on the marine ecosystem are difficult and next-to-impossible to predict and manage.

Only a few great white sharks have been documented as being killed by people in Alaskan waters and these were caught incidental to catching salmon. Since we know very little about the great white sharks found in Alaska we don't know if these catches are significant for the population; we don't even know if the great white shark population in Alaskan waters is seasonal and transitory or stable, with sharks plying secretively year around. But their secrecy may be their best strategy for their long-term survival in sub-arctic and arctic waters.

Glossary

ACUTE EFFECTS OR MORTALITIES: death that occurs shortly after exposure (e.g., oil).

ANCIENT FORESTS: forests originated through natural successions that have not experienced significant human impact.

BIODIVERSITY: the variety of life in all its forms.

BIOACCUMULATION: an increase in concentration of a compound (e.g., DDT or mercury) from the environment to the last organism in a food chain.

BIOMAGNIFICATION: an increase in concentration of a compound (e.g., DDT or mercury) from the environment from one link in a food chain to another.

BIOMASS: the amount or mass (weight) of living material.

BOTTOM-UP CONTROL: when species low on the food web, such as grazers have a significant influence on the upper levels of the food web.

CAPELIN: small, schooling fish high in calories and rich in fats (lipids).

CARNIVORES: species of animals that eat primarily meat; not to be confused with animals in the Order Carnivora, which includes carnivores and omnivores.

CASCADING EFFECT: where one species, often a keystone species, has a large-scale impact that moves up or down the food chain.

CHRONIC (HEALTH) EFFECT: an adverse health effect resulting from long-term exposure to a substance. The effects could be a skin rash, bronchitis, cancer or death.

CORRIDORS: see "landscape ecology."

DECOMPOSERS: small organisms such as molds and fungi that break down organic matter.

DELAYED IMPLANTATION: the fertilized egg will stay in a suspended state for a period of time before it implants in the uterus. It will implant only if the female is in good physical condition.

DIATOMS: a group of phytoplankton. Diatoms account for approximately 25 percent of the earth's plant matter.

ECOSYSTEM: the natural system in which energy and nutrients cycle between plants, animals and substrate.

ECOSYSTEM MANAGEMENT: the use of a holistic, long-term, ecosystem-based approach to resource management supported by ecosystem research. Ecosystem management requires consideration of geographic areas defined by ecological boundaries and of greater spatial scales and long time frames. It requires managers to take into account the complexity of natural processes and social systems and to use that understanding to craft adaptive management approaches. Ecosystem management incorporates collaborative decision-making including individuals and organizations with different values, interests and capabilities.

ECOSYSTEM RESEARCH: an all-encompassing approach to research. Many environmental parameters and trophic levels are monitored and measured to address hypotheses.

ECOTOURISM: if monitored and regulated, a relatively low-impact type of tourism that usually is focused on viewing natural phenomena.

ENDOTHERMIC: an animal that uses energy to control its internal temperature.

ESCAPEMENT (SALMON): the fish that escape being caught by fishermen and predators and which are able to spawn.

EULACHON (OR HOOLIGAN): small, schooling fish high in calories and rich in fats (lipids).

EUPHOTIC ZONE: the sunlit layer of the ocean, from the surface to the depth where only one percent of the light remains. Most primary production, growth of marine plants, takes place within the euphotic zone. The depth varies geographically and seasonally and can range from a few yards (meters) in turbid waters to around 600 feet (around 200 m) in clear waters. The ocean's average depth is 300 feet (around 100 m).

FOOD CHAIN: the predatory relationship between plants and animals. Energy and nutrients are transferred from one organism to another through the food chain.

FOOD PYRAMID: is a model of the food chain in which energy and nutrients are used to maintain each subsequent link in the food chain, and in which the amount of available energy decreases rapidly at every trophic level. Each level supports fewer individuals than does the one before. This results in a pyramid wherein there are many organisms at the bottom (producers) and few at the top (predators).

FOOD WEB: a more complex and realistic representation of a food chain that depicts the complexities of predatory relationships between plants and animals.

FORAGE FISH: a group of small, schooling fish high in calories and rich in fats. In Alaska, forage fish include herring, capelin, sandlance, and eulachon.

FURBEARERS: animals that are harvested for their furs.

GLOBAL WARMING: the warming of the earth's oceans and atmosphere; it can be man-made or natural. The present global-warming trend is being caused or enhanced by humans due to the addition of greenhouse gases, such as carbon dioxide, into the atmosphere.

GRAZERS: animals that eat plants. In the ocean there are some tiny grazers, some even microscopic, that are called zooplankton.

HABITUATED: defined in this text as when a bear becomes accustomed or used to being around people and associates people with food. When bears become habituated to people, the bears often end up being destroyed.

HERBIVORES: plant-eaters or grazers.

HERRING: small, schooling fish high in calories and rich in fats (lipids).

HIBERNATION: when a plant or animal spends the winter in a dormant state. Arctic ground squirrels hibernate (body temperature and heart rate are greatly reduced) while bears go into winter dormancy, sometimes referred to as "torpor" (body temperature drops only a few degrees and heart rate is somewhat reduced).

HOME RANGE: the area used by an animal to fulfill its needs for locating food, mates, shelter, etc.

HOT SPOT: defined in this text as an area where abundant food is found, such as a blueberry patch (terrestrial system) or a school of forage fish or salmon (aquatic system).

INDICATOR SPECIES: species that can be monitored by scientists, which reflect the health of the environment.

INTRODUCED SPECIES: a species brought by humans to an environment where it previously did not exist.

INVASIVE SPECIES: a species introduced into a new environment by humans which out-competes the natural flora or fauna and has the potential to cause environmental and economic harm.

KEYSTONE SPECIES: a species that has a large effect on the biological community despite its relative biomass.

LANDSCAPE ECOLOGY: the study of species and all the habitat necessary to the survival of the species.

LIPOPHYLLIC: having the quality of attracting lipids (fats and oils). Some contaminants (for example DDT and PCBs) are lipophyllic, which promotes their tendency to biomagnify in a food web.

MARINE PROTECTED AREA (MPA): provides protection to the flora and fauna existing with its boundaries, different MPAs providing different levels of protection. Although certain MPAs allow certain resource extraction, some MPAs restrict harvest by humans.

META-POPULATION: a collection of interacting populations of the same species.

MICROTINES: small rodents whose numbers and population densities can fluctuate dramatically from year to year (meadow mice, voles and lemmings).

MODELS: using mathematical formulas based on data to describe a system or process and make predictions.

OLD-GROWTH FORESTS: forests originated through natural successions that have not experienced significant human degradation.

OMNIVORES: animals that eat meat and plant matter. The animals in the Order Carnivora include carnivores and omnivores.

OPTIMAL FORAGING: the tendency of predators to locate and use areas of high productivity or hot spots to promote their hunting success.

OTOLITH: a small bone in the inner ear. Fish otoliths have annual rings that can be counted to determine the age of the fish.

OVOVIVIPAROUS: producing young by means of eggs hatched within the body.

PELLETS: undigested animal parts, feathers, fur and bones, formed into a cylindrical shape and regurgitated, usually by birds of prey.

PHYTOPLANKTON: small, often microscopic, free-floating oceanic plants that drift with the winds and currents.

PLANKTON: small marine plants and animals that move with the currents and winds.

POLYGAMY: a social system in which a male mates with more than one female.

PREDATOR CONTROL: human control of predator populations.

REFUGIA: protected habitat that offers some level of safety for plants or animals.

REGIME SHIFT (BIOLOGICAL): a dramatic change in the species composition of an ecosystem.

REINTRODUCTION: introducing a species back into habitat it once occupied. This technique is used to reestablish a species that has been erradicated from an area by people.

RESIDENTS (ORCAS): whales that do not leave their pod for extended periods of time, are more vocal, have larger pods (five to 50 animals), and eat only or mostly fish.

SANDLANCE: small, schooling fish high in calories and rich in fats (lipids). The nickname is "needlefish."

SATELLITE TAGS: electronic units that are attached to animals, which send data to satellites which then send the data to scientists so that animal movement can be tracked.

TERRITORY: the part of the home range that is defended against intrusion.

TERRESTRIAL: on land.

THERMOCLINE: an abrupt change in water temperature in the water column.

TOP-DOWN CONTROL: when species high in the food web, such as predators, have a significant influence on the lower levels of the food web.

TRANSIENTS (ORCAS): orcas that have a flexible social structure in that animals may leave the pod for many years. Compared to resident orcas, transient orcas are less vocal and more secretive, travel in smaller pods, usually two to seven individuals, and eat only or mostly mammals.

TRICHINOSIS: a disease caused by eating meat that contains infective *Trichinella* cysts. In the host's stomach the cysts hatch into small parasitic worms.

Information Sources

Alaska has thousands of predatory species, so we could not possibly include them all in this book. We would like to have had chapters on northern pike, arrowtooth flounder, all five species of salmon, peregrine falcon, great-gray owl, red fox, lingcod, northern fur seal, humpback whales, jellyfish, warblers, spiders, bats, and the list goes on. We selected what we think are representative species and species that we have studied, but all the predators in Alaska play an important role in the ecosystem and are of equal value and interest. We hope this book motivates you to learn more about all the predators.

Much of the information found in this book was derived from scientific literature (peer-reviewed journals), attending scientific meetings where predator information was presented, and from meetings and discussions with individual scientists. The author's years of field observations are another important source of information. He has spent much time in Alaska's wilderness and on Alaska's waterways studying Alaska's wildlife and ecosystems.

TO LEARN MORE ABOUT CONSERVATION BIOLOGY READ: *Principles of Conservation Biology,* 2nd edition. Gary Meffe and Ronald Carroll. 1997. Sinauer Associates, Inc.

TO LEARN MORE ABOUT FOREST PREDATORS READ: *The Scientific Basis for Conserving Forest Carnivores: American Marten, Fisher, Lynx and Wolverine in the Western United States.* L.F. Ruggiero, K.B. Aubry, S.W. Buskirk, L.J. Lyons and W.J. Zielinski. 1994. U.S. Department of Agriculture, Forest Service, Rocky Mountain Research Station, Fort Collins. GTR-RM-254.

Wolves, bears, and their prey in Alaska: Biological and social challenges in wildlife management. National Research Council. 1997. National Academy Press.

TO LEAN MORE ABOUT MAMMALS READ: *Mammals of North America: Temperate and Arctic Regions.* Adrian Forsyth. 1999. Firefly Books Ltd.

TO LEARN MORE ABOUT THE *EXXON VALDEZ* OIL SPILL READ: *Sound truth and corporate myth$. The legacy of the Exxon Valdez oil spill.* R. Ott. 2004. Dragonfly Sisters Press.

AN INTERESTING PREDATOR-PREY PAPER: "Killer whale predation on sea otters linking oceanic and nearshore ecosystems." J.A. Estes, M T. Tinker, T.M. Williams and D.F. Doak. 1998. *Science,* 282, 473–476.

BRUCE WRIGHT IS STANDING ON A PEAK IN THE TALKEETNA MOUNTAINS WITH DENALI IN THE BACKGROUND.
PHOTO BY SADIE WRIGHT

About the Author

Bruce Wright earned an advanced degree in ecology studying birds of prey at San Diego State University, and he was a professor at the University of Alaska where he taught courses about bald eagles, orcas, humpback whales, and brown, black and polar bears. He also was the president of the board of directors of the Bald Eagle Research Institute.

When Bruce was a section chief for the National Oceanic and Atmospheric Administration (NOAA) he continued his work with predators in Alaska by managing the Alaska Predator Ecosystem Experiment and the Alaska Shark Assessment Project. He was actually bitten by one of the 350-pound sharks that his crew landed during the field research. Bruce was selected as the chief science advisor to Alaska's Governor Tony Knowles when the governor worked with the Pew Ocean Commission on ocean issues.

Bruce is the executive director of the Conservation Science Institute (conservationinstitute.org), senior scientist to the Aleutian Pribilof Islands Association, and principal investigator to a Bering Sea contaminants-monitoring project and to another project designed to monitor paralytic shellfish poisoning in the Aleut region of the Alaska Peninsula, including the Aleutian and Pribilof Islands of Alaska and the Commander Islands in Russia.

INDEX

A

abalone 74
acute mortalities 93
aircraft 60, 102, 103
Alaska Department of Fish and Game 36, 43, 78
Alaska Trappers Association 62
Aleutian Canada goose 28, 48
Aleutian Islands 8, 25, 26, 27, 28, 29, 38, 47, 57, 93, 107, 110, 116
Alopex lagopus 25
ancient forests 60, 61, 112
arctic fox 5, 9, 23, 24, 25, 26, 27, 28, 101
arctic ground squirrels 27, 36, 113
arrowtooth flounder 80, 82, 85, 96, 115

B

bachelor groups 91, 92
Bald and Golden Eagle Protection Act 61
bald eagle 5, 7, 8, 9, 13, 18, 23, 48, 57, 58, 59, 60, 61, 62, 65, 81, 91, 92, 116
baleen whale 105
Barren Islands 21
bear 5, 7, 9, 12, 18, 20, 23, 27, 31, 32, 33, 34, 35, 36, 37, 38, 39, 40, 41, 42, 43, 51, 83, 98, 99, 100, 101, 102, 103, 113
bear scat 35, 36
bends 104
Berners Bay 59
big game 61
biodiversity 15, 112
biomagnification 61, 103, 107, 109, 112
biomass 13, 14, 31, 112, 113
black bear 5, 9, 23, 32, 33, 34, 35, 36, 42

boars 32, 33, 36, 37, 38
boom and bust 17
bottom-up control 5, 14, 79, 112
bounty 60
brown bear 5, 9, 18, 20, 23, 33, 35, 36, 37, 38, 39, 40, 41, 42, 43, 98, 101
Bubo virginianus 65

C

camouflaged 17, 79, 83, 87, 88, 103, 109
Canidae 23, 25, 28
Canis lupus 28
capelin 21, 80, 82, 85, 86, 88, 95, 96, 112, 113
caribou 15, 26, 28, 30, 31, 32, 35, 39, 44, 52
Carnivora 32, 112, 114
carnivores 14, 15, 30, 32, 48, 74, 98, 112, 114, 115
cascading effect 78, 112
channels 80, 106
Chichagof Island 38, 41
chronic effects 93, 112
clams 34, 39, 74, 75, 76, 77, 78, 92, 93
clear-cut logging 22, 51
community structure 14
consumers 13, 15
contaminants 14, 22, 60, 61, 62, 65, 68, 78, 90, 93, 96, 103, 107, 113, 116
contaminated prey 93, 103
Copper River Delta 59
corridors 20, 34, 39, 112
countercurrent heat exchange 80
coupled relationship 17
coyotes 7, 12, 17, 27, 28
cubs 7, 32, 33, 35, 36, 37, 38, 98, 99, 102

D

Dall's porpoise 7, 107
DDT 14, 61, 65, 103, 108, 112, 113
decomposers 13, 112
deer 15, 30, 32, 34, 35, 39, 59, 70
delayed implantation 33, 37, 46, 48, 51, 54, 91, 94, 98, 112
diatoms 85, 112

E

ecosystem 8, 12, 13, 14, 15, 16, 18, 20, 21, 23, 32, 36, 65, 68, 75, 78, 79, 82, 85, 86, 90, 93, 97, 107, 111, 112, 114, 115, 116
ecosystem management 12, 16, 112
ecosystem research 12, 112
ecotourism 28, 65, 109, 112
electromagnetic fields 80, 83
Endangered Species Act 96
endothermic 79, 109, 112
Enhydra lutris 90
escapement 15, 112
eulachon 52, 59, 95, 112, 113
Eumetopias jubatus 94
euphotic zone 21, 74, 77, 78, 85, 87, 112
extinct 86, 97
Exxon Valdez 61, 93, 109, 115

F

Falco rusticolus 62
food chains 14
food pyramid 14, 113
food web 13, 14, 61, 65, 85, 93, 95, 103, 105, 107, 112, 113, 114
forage fish 16, 20, 21, 59, 82, 85, 86, 94, 95, 96, 97, 113
fox farms 60
foxes 8, 12, 19, 25, 26, 27, 28, 29, 44, 48, 50, 60, 65, 71, 101
furbearers 61, 113

G

gene flow 5, 9, 19, 20
geography 12, 21
geology 21
giant Pacific octopus 5, 23, 76, 77, 78
glacier bear 32
global warming 16, 22, 85, 86, 90, 103, 113
Gobi Desert 85
golden eagles 62, 65
grazers 78, 93, 112, 113
great horned owl 5, 9, 21, 23, 65, 66, 67, 68
grizzly bears 7, 27, 34, 36, 65
Gulo gulo 51
guns 60
gyrfalcon 5, 9, 13, 23, 62, 63, 64, 65

H

habituated 36, 41, 113
Haliaeetus leucocephalus 57
hare 17, 18, 21, 27, 28, 31, 44, 45, 47, 50, 59, 64, 66, 70
herbivores 13, 14, 15, 30, 113
herring 20, 21, 59, 79, 80, 82, 85, 86, 88, 95, 96, 105, 113
hibernate 27, 36, 98, 113
hibernation 36, 113
Hippoglossus stenolepis 86
home range 5, 9, 18, 44, 47, 49, 50, 52, 88, 98, 105, 113, 114
hooligan 112
horned owls 7, 18, 50, 65, 66, 67, 68
hot spots 5, 9, 19, 20, 21, 79, 80, 82, 94, 105, 111, 114
human persecution 62, 68
hunters 27, 43, 44, 50, 61
hypotheses 17, 112

I

indicator species 5, 9, 12, 14, 22, 61, 113
indicators 23
insecticides 68
International Pacific Halibut Commission 83, 88, 90
interspecific competition 5, 9, 21
intertidal zone 47, 74, 75, 77, 78, 95
introduced 25, 27, 28, 48, 49, 51, 78, 113
Inuit 65
invasive species 9, 22, 51, 78, 79, 113
iron 15, 85

K

Kenai Peninsula 40, 59
keystone predator 90
keystone species 5, 9, 14, 15, 16, 31, 79, 112, 113
killer whale 9, 103, 115
Kootznoowoo 38

L

Lamna ditropis 79
Lamnidae 23, 79, 109
landscape 5, 9, 13, 18, 19, 20, 21, 34, 38, 39, 112, 113
landscape ecology 5, 9, 19, 112, 113
lead 8, 61
lead poisoning 61
lemming 20, 27, 28, 71
lipophyllic 61, 113
loss of habitat 22, 68
Lower Cook Inlet 21
lynx 5, 9, 17, 23, 28, 32, 43, 44, 45, 48, 50, 115
Lynx canadensis 43

M

Marine Mammal Protection Act 102 3
marine parks 107
marine protected areas 90
marine reserves 22, 90
marten 5, 8, 9, 19, 20, 23, 45, 46, 48, 49, 50, 51, 61, 66, 90, 115
Martes americana 48
mathematical models 17, 82
matrix 19, 20
mercury 90, 112
meso-scale 12
microlayer 78
microtine 44
mink 5, 8, 9, 19, 23, 45, 46, 47, 48, 50, 51, 61, 90
models 17, 82, 86, 113
moose 8, 15, 30, 31, 32, 34, 35, 39, 49, 52, 106
mud shark 82
mussels 39, 74, 75, 76, 92, 93
Mustela vison 45
Mustelidae 23, 45, 48, 51

N

niche 5, 9, 21, 22
niche separation 21
non-target animals 62
northern sea lion 94
Nyctea scandiaca 69

O

odontoceti 103
offshore whales 104
oil and gas development 103
oil pollution 93
oil spills 61, 93, 103, 109
old growth forests 61
omnivores 14, 32, 112, 114
optimal foraging 5, 17, 114
Orcinus orca 103

Oregon 59, 91
osprey 61
otolith 83, 87, 114
over-fishing 22
ovoviviparous 79, 114

P

Pacific halibut 5, 9, 23, 83, 86, 87, 88, 89, 90
Pacific sleeper shark 5, 9, 23, 82, 83, 84, 85
pack ice 98, 105
paralarvae 77
patch 7, 20, 21, 82, 98, 113
pellets 67, 68, 71, 114
phytoplankton 14, 15, 16, 21, 78, 85, 112, 114
phytoplankton bloom 21
pigs 32
planktonic 87, 88
pods 92, 95, 104, 105, 106, 107, 109, 114
poisoning 61, 68, 93, 116
polar bear 5, 9, 12, 23, 27, 98, 99, 100, 101, 102, 103
polygamous 104
power lines 60
precautionary approach 82, 90
predator control 5, 9, 15, 32, 114
prey availability 12, 14, 44, 52, 59, 105
Prince William Sound 8, 48, 61, 73, 79, 81, 107, 108, 109, 110
producers 13, 14, 113
ptarmigan 18, 21, 28, 44, 52, 64, 70, 71
pups 22, 27, 29, 79, 81, 91, 94, 101
Pycnopodia helianthoides 73

R

raptor rehabilitation centers 61
red fox 7, 25, 26, 115
refugia 55, 114

regime shift 16, 20, 79, 85, 90, 96, 97, 114
rehabilitation centers 61, 93
reintroduction 62, 92, 114
relocating 36
residents 66, 103, 104, 106, 114
ringed seal 101
Russian fur traders 92

S

salmon 5, 7, 8, 9, 15, 16, 20, 21, 23, 29, 31, 33, 34, 35, 36, 37, 38, 39, 40, 41, 49, 50, 52, 58, 59, 60, 79, 80, 81, 82, 83, 85, 86, 88, 94, 95, 96, 105, 106, 109, 110, 111, 112, 113, 115
salmon shark 5, 9, 20, 21, 23, 79, 80, 81, 82
San Francisco Bay 79
sandlance 85, 86, 88, 95, 96, 113, 114
satellite tags 79, 83, 110, 114
scent marking 18, 49
sea otter 5, 9, 23, 48, 81, 90, 91, 92, 93, 107
sea urchin 14, 75, 93, 107
shark 5, 7, 9, 20, 21, 22, 23, 79, 80, 81, 82, 83, 84, 85, 86, 109, 110, 111, 116
single-species management 12
Sitka Sound 59
sixth sense 80, 83
snowy owl 5, 9, 23, 28, 69, 70, 71
Somniosus pacificus 82
species diversity 13
Steller's sea lion 5, 9, 23, 94, 96
Stikine River 59
Strigiformes 65, 69
sunflower star 5, 9, 17, 23, 73, 74, 75

T

talons 19, 60, 66, 67, 69

territory 5, 9, 18, 19, 21, 29, 49, 51, 57, 66, 69, 71, 94, 114
thermocline 80, 114
timber 41, 60, 61
tools 92
top-down control 14, 31, 107, 114
transients 49, 93, 104, 105, 106, 114
trapping 17, 27, 30, 31, 45, 48, 50, 51, 54, 55, 60, 61, 71
trichinosis 32, 36, 43, 114
trophic cascade 75, 107
tube feet 74, 75

U

U.S. Fish and Wildlife Service 61
U.S. Navy 79
ungulate 31
upwelling 21, 105
Ursus americanas 32
Ursus arctos 36
Ursus maritimus 98

W

Washington 59, 91
whale carcass 102
whelps 25, 26, 27
wolverine 5, 9, 12, 18, 19, 23, 28, 32, 45, 48, 51, 52, 53, 54, 55, 90, 115

Z

zooplankton bloom 21

Other Hancock House natural history titles

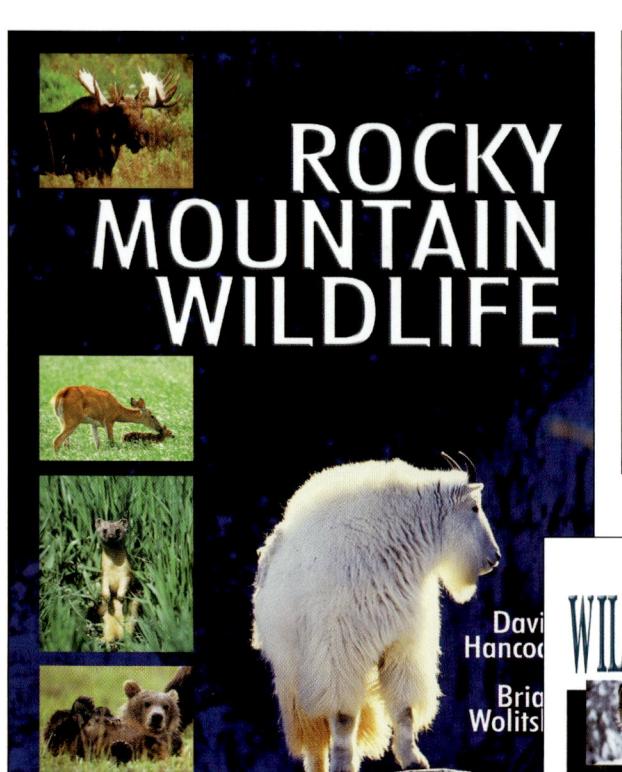

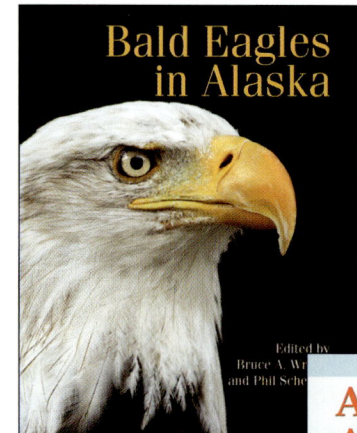

Bald Eagles in Alaska
Bruce Wright & Phil Schempf, Editors
978-0-88839-590-0
8½ x 11, sc, 438 pp

Attending Alaska's Birds
James G. (Jim) King
978-0-88839-656-3
7 x 10, sc, 480 pp

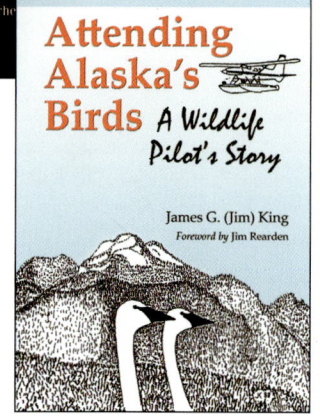

Rocky Mountain Wildlife
David Hancock & Brian Wolitski
978-0-88839-567-2
8½ x 11, sc, 96 pp

Wildlife of the North
Steven Kazlowski
978-0-88839-590-0
8½ x 11, sc, 48 pp

The Bald Eagle of Alaska, BC & WA
David Hancock
978-0-88839-536-8
5½ x 8½, sc, 96 pp

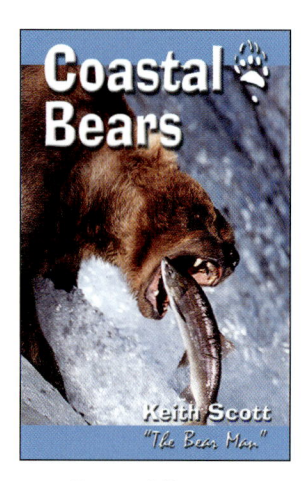

Coastal Bears
Keith Scott
978-0-88839-626-6
5½ x 8½, sc, 80 pp

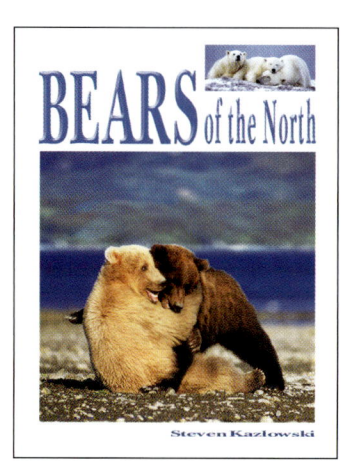

Bears of the North
Steven Kazlowski
978-0-88839-591-7
8½ x 11, sc, 96 pp

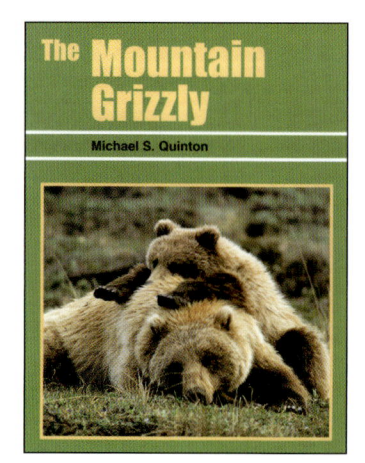

The Mountain Grizzly
Michael S. Quinton
978-0-88839-625-9
8½ x 11, sc, 64 pp

Hancock House Publishers

www.hancockhouse.com